Engineering English
for Mechanical Engineers

D.H.KIM · S.S.YOU · G.S.LEE · D.H.DOH

도서출판 GS인터비전

머리말

P·R·E·F·A·C·E

대학사회에 국제화의 물결이 몰려오기 시작한 것도 벌써 10년이 넘었습니다. 이는 1988년 서울올림픽을 계기로 우리나라가 세계무대에 진출한 이후 사회 각 분야의 세계화를 추진하는 과정에서 어쩔 수 없는 선택이라고 보여 집니다. 이에 따라 대학을 졸업하는 학생들의 취업전선에는 외국어 구사능력 확보라는 또 하나의 짐이 추가되었고 가장 보편적인 세계 공용어인 영어를 제대로 구사할 수 있는 졸업생이 취업에 있어서 유리한 고지를 점유할 수 있게 되었습니다. 일류 대기업에 취업하기 위해서는 적어도 800점 이상의 TOEIC 점수를 받아야 하고 경우에 따라서는 900점 이상을 요구하기도 합니다.

평균적으로 볼 때 이공계 대학생들은 인문사회계 학생들보다 영어 구사 능력이 떨어집니다. 이는 우리나라 고교 교육에서부터 발원된 문제로 대학에 들어와서도 전공에 중점을 두다보면 이러한 영어 구사 능력에 있어서의 실력차가 줄어들지 않고 오히려 벌어지는 경향이 있는 것 또한 사실입니다. 그러나 이공계에서 사용하는 영어의 수준이나 패턴은 인문사회계에서 사용하는 것과 비교해볼 때 비교적 단순한 면이 있기 때문에 이공계 대학생들이 조금만 노력한다면 쉽게 영어를 이해하고 구사할 수 있습니다.

이공계 대학생들이 졸업 전에 습득해야할 영어 중에서 가장 중요한 부분이 전공과 관련된 영어입니다. 우리나라에서의 공학교육이 서구에 비해 짧고 또한 공학 자체가 서구에서 유래된 관계로 공학 용어 중 상당부분이 제대로 번역되지 못하고 있습니다. 또한 우리말로 제대로 변역된 용어들도 국제화가 가속화 될수록 실제의 산업현장에서 잊혀져가고 오히려 원어로 된 용어가 점점 더 많이 사용되고 있는 게 현실입니다.

근래 들어 점점 번역본이 많이 사용되는 추세인 점을 감안한다면 대학에서 원어로 된 용어를 별도로 배우지 않고서는 국제화되어가는 기업체의 요구를 충족시키기 어렵게 되었습니다. 다행히 많은 공과대학에서 전공영어 혹은 공학영어라는 이름으로 과목을 개설하여 학생들을 교육시키고는 있으나 우리 실정에 맞는 교재의 부재로 인하여 적절한 교육에 애로를 겪고 있는 것으로 알고 있습니다.

이에 저희들은 최소한 기계공학의 범주에 드는 학과들에 맞는 교재를 편찬하기로 결정하고 이 분야에서 다루는 대부분의 과목에 대한 영어원서를 망라하여 학부생들에게 적절한 수준의 전공 예문들을 엄선하였습니다. 이들 예문들을 철저히 공부함으로써 학생들은 전공에서 사용하는 전문용어를 원어로 접할 수 있게 됨은 물론이고 원서의 해독에 자신감을 가질 수 있게 될 것입니다. 또한 선정된 전공 예문을 습득함으로써 전공 일반에 대한 지식을 두루 쌓을 수 있는 부수적인 효과도 누리게 될 것입니다.

본 교재는 모두 16개의 유닛으로 구성되어 1학기 동안 배울 수 있는 분량으로 설정되었습니다. 실제로 두 번의 평가시험 기간을 뺀다면 13개 정도의 유닛을 다룰 수 있을 것이므로 필요에 따라 중점을 두고자 하는 유닛들로 한 학기 동안의 강의 내용을 구성할 수 있을 것입니다.

각 유닛에는 문장의 이해도를 측정하기 위해 연습문제들을 삽입하였으며 학생들이 이 들을 과제물로 하여 성실히 풀어본다면 강의를 통해 배웠던 내용들을 잘 이해할 수 있을 것으로 보입니다.

아무쪼록 본 교재를 잘 활용하여 여러분의 전공 지식 함양은 물론 영어를 통한 전문용어 및 원서 해독능력 습득 그리고 나아가 TOEIC 성적 향상에 있어서 성과를 거두기 바랍니다.

2008. 2

저자 일동

차 례

C·O·N·T·E·N·T·S

❖ Unit 05 | Fluid Power 67

❖ Unit 06 | Energy 85

❖ Unit 07 | Thermal Engineering 97

❖ Unit 08 | Mechanisms 113

❖ Unit 09 | Manufacturing Processes and Mechanical Vibrations 127

❖ Unit 15 | Industrial Robot *223*

❖ Unit 16 | Mechanical Design and MEMS *239*

Unit

01

Mechanical Engineering

Unit 01 Mechanical Engineering

1 Mechanical Engineers

Mechanical engineering deals with power and the design of machines and processes used to generate power and apply it to useful purposes. These designs may be simple or complex, inexpensive or expensive, luxuries or essentials. Such items as the kitchen food mixer, the automobile, air-conditioning systems, nuclear power plants, and interplanetary space vehicles would not be available for man's use today were it not for the mechanical engineer. In general, the mechanical engineer works with systems, subsystems, and components that have motion. The range of work that may be classed as mechanical engineering is wider than that in any of the other branches of engineering, but it may be grouped generally under two heads: work that is concerned with power-generating machines, and work that deals with machines that transform or consume this power in accomplishing their particular tasks. The utilization of solar energy for domestic and industrial uses is one of the more important areas of the mechanical engineer's work at this time.

Some of the general subdivisions of mechanical engineering are as follows: Power or combustion engineers deal with the production of power from fuels. Design specialists may work with parts that vary in size from the microscopic part of the most delicate instrument to the massive part of heavy machinery. Included are the mass transit systems that are rapidly

Terminology

air–conditioning
combustion engineer
heavy machinery
interplanetary space
nuclear power plant
solar energy
transit system

becoming a part of heavy machinery. Automotive engineers work constantly to improve the vehicles and engines that we now have. Heating, ventilating, air-conditioning, and refrigeration engineers deal with the design of suitable systems for making our building more comfortable and for providing proper conditions in industry for good working conditions and efficient machine operation.

Employment may be secured by mechanical engineering graduates in almost every type of industry. Manufacturing plants, power-generating stations, public utility companies, transportation companies, airlines, and factories, to mention only a few, are examples of organizations that need mechanical engineers. Experienced engineers are needed in the missile and space industries in the design and development of such items as gas turbine compressors and power plants, air-cycle cooling turbines, electrically and hydraulically driven fans, and high-pressure refrigerants. Mechanical engineers are also needed in the testing of airborne and missile fuel systems, servovalves, and mechanical-electrical control systems. In addition, an engineer may be employed for research endeavor as a university professor, or in the governments of cities, states, and the nation.

Terminology

airborne
compressor
gas turbine
public utility
refrigeration
servovalves
ventilating

EXERCIS 1-1

A. Words and Phrases

1. deal with	2. luxury	3. be available for
4. consume	5. utilization	6. microscopic
7. delicate	8. massive	9. constantly
10. to mention only a few	11. hydraulically	12. in addition
13. be employed for	14. endeavor	

B. Reading Comprehension

1. Put T or F to indicate if the sentences below are true or false according to the facts in the passage.

- Mass transit systems are not included in the general subdivisions of mechanical engineering. ________________
- The mechanical engineer deals with systems, subsystems, and components that have motion. ________________
- Mechanical engineering deals with power and the design of machines and processes. ________________
- Employment for mechanical engineering graduates is not spread widely. ________________
- The missile and space industries do not need mechanical engineers. ________________
- The utilization of solar energy belongs to mechanical engineering. ________________

2. Answer the following questions.

- Group mechanical engineering under two heads.
- List up the possible employment chances for mechanical engineering graduates.

bibliography

George C. Beakley and H.W. Leach, *Engineering: An Introduction to Creative Profession*, pp. 104–105, Macmillan Publishing Co, 1977.

2 Engineering Mechanics

The branch of scientific analysis which deals with motions, time, and forces is called mechanics and is made up of two parts, statics and dynamics. Statics deals with analysis of stationary systems, i.e., those in which time is not a factor, and dynamics deals with systems which change with time.

As shown in Fig 1.1, dynamics is also made up of two major discipline, first recognized as separate entities by Euler in 1775.

These two aspects of dynamics were later recognized as the distinct sciences of kinematics (from the Greek word kinema, meaning motion) and kinetics, and deal with motion and the forces producing it, respectively.

The initial problem in the design of mechanical system is therefore understanding its kinematics. Kinematics is the study of motion, quite apart from the forces which produce that motion. More particularly, kinematics is the study of position, displacement, rotation, speed, velocity, and acceleration. The study, say, of planetary or orbital motion is also a problems which arise in the design of mechanical systems.

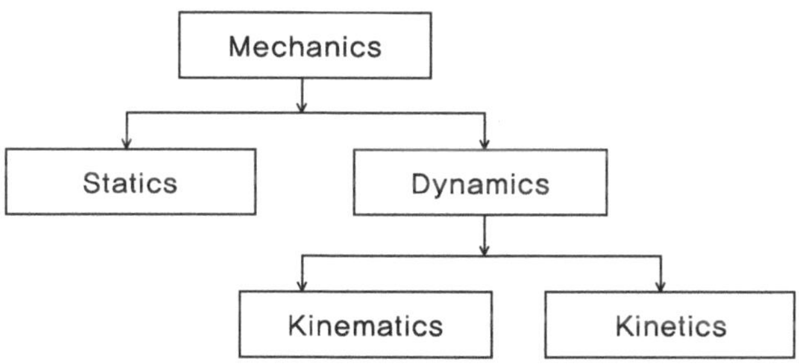

▲ Figure 1.1

Terminology

acceleration
kinetics
orbital motion
kinematics
velocity
mechanics
statics
dynamics

It should be carefully noted in the above quotation that Euler based his separation of dynamics into kinematics and kinetics on the assumption that they should deal with rigid bodies. It is this very important assumption that allows the two to be treated separately. For flexible bodies, the shapes of the bodies themselves, and therefore their motions, depend on the forces exerted on them. In this situation, the study of force and motion must take place simultaneously, thus significantly increasing the complexity of the analysis.

Fortunately, although all real machine parts are flexible to some degree, machines are usually designed from relatively rigid materials, keeping part deflections to a minimum. Therefore, it is common practice to *assume* that deflections are negligible and parts are rigid when analyzing a machine's kinematic performance, and then, after the dynamic analysis when loads are known, to design the parts so that this assumption is justified.

Terminology

dynamic analysis
flexible body
rigid body

EXERCIS 1-2

A. Words and Phrases

1. branch
2. be made up of
3. stationary
4. discipline
5. apart from
6. negligible
7. quotation
8. assumption
9. rigid body
10. simultaneously
11. significantly
12. complexity
13. relatively
14. common practice
15. deflection

B. Reading Comprehension

1. Put T or F to indicate if the sentences below are true or false according to the facts in the passage.

- Mechanics deals with motions, time, and forces. ________
- Dynamics is divided into kinematics and kinetics. ________
- Kinematics is the study of position, displacement, rotation, and so on. ________
- Deflections can not be neglected when analyzing a machine's kinematic performance. ________
- All real machine parts are flexible to some degree. ________
- Kinematics involves the forces which produce motion. ________
- Euler's separation of dynamics into kinematics and kinetics is based on the assumption that they should deal with rigid bodies. ________

2. Answer the following questions.

- Write down the differences between kinematics and kinetics.
- List the two parts which make up mechanics.
- What is Euler's assumption on which his separation of dynamics into kinematics and kinetics is based?

bibliography

Joseph Edward Shigley and Joseph Uiker JR, *Theory of Machines and Mechanisms*, pp. 4~6, McGraw-Hill, 1995.

3) Newton's Laws of Motion

Many of the earlier notions about dynamics were dispelled after 1590 when Galileo performed experiments to study the motions of pendulums and falling bodies. The conclusions drawn from these experiments gave some insight as to the effects of forces acting on bodies in motion. The general laws of motion of a body subjected to forces were not known, however, until 1687, when Isaac Newton first presented three basic laws governing the motion of a particle. In a slightly reworded form, Newton's three laws of motion can be stated as follows:

First Law: A particle originally at rest, or moving in a straight line with a constant velocity, will remain in this state provided the particle is not subjected to an unbalanced force.

Second Law: A particle acted upon by an unbalanced force **F** experiences an acceleration a that has the same direction as the force and a magnitude that is directly proportional to the force.

Third Law: The mutual forces of action and reaction between two particles are equal, opposite, collinear.

Newton's three laws were developed from a study of planetary motion (the motion of particles). During the eighteenth century, Leon-hard Euler (1707-1783) extended Newton's work on particles to rigid-body systems.

Newton's first law is known as the principle of inertia. Inertia is the property of matter that makes an object stay in motion if it is moving or remain motionless if it is not moving. Forces that change an object's motion must first overcome inertia of the object. The greater an object's mass, the harder it is to put the object into motion or change its velocity. Inertia serves as a measurement of how hard it is to get an object moving.

Terminology

inertia
particle
straight line
pendulum
collinear
unbalanced force

Newton's second law may be written as **F** = (mass) x (acceleration) = m**a**, where **F** is the applied force. Scientists use this formula to describe motion of all kinds of objects.

For example of Newton's third law, when hot gases escape from a rocket engine during take-off, the rocket is propelled upward. The downward motion of the gases from the rocket generates a reaction of the rocket upward. This reaction helps the rocket overcome air resistance and fly into space. There are many other examples of Newton's laws.

EXERCIS 1-3

A. Words and Phrases

1. dispel
2. pendulum
3. effect
4. be subjected to
5. reworded form
6. at rest
7. be proportional to
8. mutual
9. reaction
10. according to
11. escape from
12. take-off
13. propel
14. upward
15. downward

B. Reading Comprehension

1. Put T or F to indicate if the sentences below are true or false according to the facts in the passage.

- A study of planetary motion made Newton develop three laws. ________
- The principle of inertia is the Newton's third law. ________
- Galileo's experiments to study the motion of falling bodies influenced Newton. ________
- Isaac Newton presented three basic laws of motion in 1687. ________
- Euler extended Newton's work to particles. ________
- The formula to describe motion of all kinds of objects is the Newton's first law. ________

2. Answer the following questions.

- List the Newton's three basic laws and explain in detail.
- Who extended Newton's work on particles to rigid-body systems?

bibliography

R.C. Hibbeler, *Engineering Mechanics(Dynamics)*, pp. 97~98, Prentice-Hall, 1998.

Unit 02

Heat and Energy

Unit 02 Heat and Energy

1 Thermodynamics and Energy

Thermodynamics can be defined as the science of energy. Although everybody has feeling of what energy is, it is difficult to give a precise definition for it. Energy can be viewed as the capacity to do work or as the ability to cause changes.

One of the most fundamental laws of nature is the **conservation of energy** principle. It simply states that during an interaction energy can change from one form to another but that the total amount of energy remains constant. That is, energy cannot be created or destroyed. A rock falling off a cliff, for example, picks up speed as a result of its potential energy being converted to kinetic energy. (Fig 2.1) The conservation of energy principle also forms the backbone of the diet industry: a person who has a greater energy input (food) than energy output (exercise) will gain weight(store energy in the form of fat), and a person who has a smaller energy input than output will lose weight. (Fig 2.2).

Terminology

conservation of energy
thermodynamics
potential energy
kinetic energy

Thermodynamics deals with the conservation of energy from one form to another. It also deals with various properties of substances and the changes in these properties as a result of energy transformations. Like all sciences, thermodynamics is based on experimental observations. The findings from these observations have been expressed as some basic laws. **The first law of thermodynamics**, for example, is simply an expression of the conservation of energy principle. The **second law of thermodynamics** asserts that processes occur in a certain direction but not in the reverse direction. A cup of hot coffee left on a table in an office, for example, eventually cools, but a cup of cool coffee on the same table never gets hot by itself. (Fig 2.3)

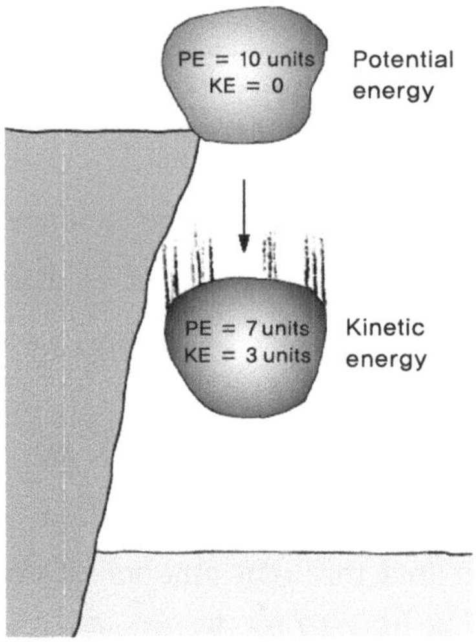

▲ **Figure 2.1** Energy cannot be created or destroyed; it can only change forms(the first law).

▲ **Figure 2.2** Conservation of energy principle for the human body.

▲ **Figure 2.3** Heat can flow only from hot to cold bodies (the second law).

Although the principles of thermodynamics have been in existence since the creation of the universe, thermodynamics did not emerge as a science until about 1700 when the first attempts to build a steam engine were made in England by T. Savery and T. Newcomen. These engines were very slow and inefficient, but they opened the way for the development of a new science. The term thermodynamics was first used in a publication by Lord Kelvin in 1849. The first thermodynamics textbook was written in 1859 by W. Rankine, a professor at the University of Glasgow. The greatest progress in thermodynamics was made in the early 1890s when it was stripped of all the erroneous theories and developed into a mature science.

It is well known that a substance consists of a large number of particles called molecules. The properties of the substance naturally depend on the behavior of these particles. For example, the pressure of a gas in a container is the result of momentum transfer between the molecules and the walls of the container. But one does not need to know the behavior of the gas particles to determine the pressure in the container. It would be sufficient to attach a pressure gage to the container. This macroscopic approach to the study of thermodynamics which does not require a knowledge of the behavior of individual particles is called **classical thermodynamics**. It provides a direct and easy way to the solution of engineering problems. A more elaborate approach, based on the average

Terminology

macroscopic
momentum transfer
molecule
steam engine
classical thermodynamics

behavior of large groups of individual particles, is called **statistical thermodynamics**. This microscopic approach is rather involved and is used in this text only in the supporting role.

Application Areas of Thermodynamics

Every engineering activity involves an interaction between energy and matter, thus it is hard to imagine an area which does not relate to thermodynamics in some respect. Therefore, developing a good understanding of thermodynamics principles has long been an essential part of engineering education.

One does not need to go very far to see some application areas of thermodynamics. In fact, one does not need to go anywhere. These areas are right where one lives. An ordinary house is, in some respects, an exhibition hall filled with thermodynamic wonders. Many ordinary household utensils and appliances are designed, in whole or in part, by using the principles of thermodynamics. Some examples include the electric or gas range, the heating and air-conditioning systems, the refrigerator,

Terminology

air-conditioning
refrigerator
microscopic
statistical thermodynamics

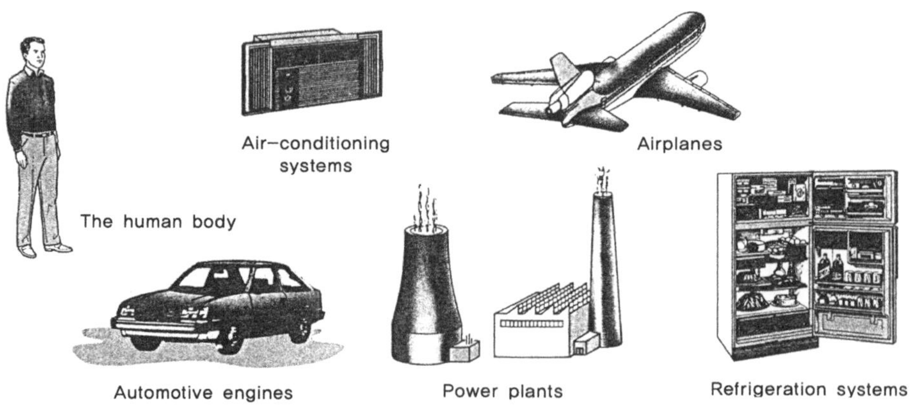

▲ **Figure 2.4** Some application areas of thermodynamics.

the humidifier, the pressure cooker, the water heater, the shower, the iron, and even the computer, the TV, and the VCR set. On a larger scale, thermodynamics plays a major part in the design and analysis of automotive engines, rockets, jet engines, and conventional or nuclear power plants. (Fig 2.4) We should also mention the human body as an interesting application area of thermodynamics.

Terminology

humidifier
nuclear power
jet engine

EXERCIS 2-1

A. Words and Phrases

1. falling off	2. cliff	3. backbone
4. assert	5. reverse	6. emerge
7. erroneous	8. mature	9. elaborate
10. exhibition	11. utensil	12. appliance

B. Reading Comprehension

1. Put T or F to indicate if the sentences below are true or false according to the facts in the passage.

- The principles of thermodynamics emerged in the 18th century. ________________
- The conservation of energy principle is the basis of the first law of thermodynamics. ________________
- Thermodynamics is based on theories rather than experiments. ________________
- The second law of thermodynamics asserts that processes occur in both directions. ________________
- Lord Kelvin used the term thermodynamics in a conference meeting. ________________
- The macroscopic approach to the study of thermodynamics is called statistical thermodynamics. ________________
- The first thermodynamics textbook was written by a professor at the University of Glasgow. ________________

- The behavior of molecules determines the properties of the substance. ________________

2. Answer the following questions.

- How can thermodynamics be defined in this passage?
- What does the conservation of energy principle simply state according to this passage?
- When did thermodynamics emerge as a science?
- Explain classical thermodynamics and statistical thermodynamics, respectively.
- List some examples of the ordinary household utensils and appliances which use the principles of thermodynamics.

bibliography

Yuns A. Cengel and Michael A. Boles, *Thermodynamics: An Engineering Approach*, pp. 2~3, McGraw-Hill, 1989.

2 The Relation of Heat Transfer to Thermodynamics

Whenever a temperature gradient exists within a system, or whenever two systems at different temperatures are brought into contact, energy is transferred. The process by which the energy transport takes place is known as heat transfer.

The thing in transit, called heat, cannot be measured or observed directly, but the effects it produces are amenable to observation and measurement. The flow of heat, like the performance of work, is a process by which the internal energy of a system is changed.

The branch of science that deals with the relation between heat and other forms of energy is called thermodynamics. Its principles, like all laws of nature, are based on observations and have been generalized into laws that are believed to hold for all processes occurring in nature because no exceptions have ever been found. The first of these principles, the first law of thermodynamics, states that energy can be neither created nor destroyed but only changed from one form to another. It governs all energy transformations quantitatively but places no restrictions on the direction of the transformation. It is known, however, from experience that no process is possible whose sole result is the net transfer of heat from a region of lower temperature to a region of higher temperature. This statement of experimental truth is known as the second law of thermodynamics.

All heat transfer processes involve the transfer and conversion of energy. They must therefore obey the first as well as the second law of thermodynamics. At first glance one might therefore be tempted to assume that the principles of heat transfer can be derived from the basic laws of thermodynamics. This conclusion, however, would be erroneous, because classical thermodynamics is restricted primarily to the study of equilibrium states, including mechanical and chemical as well as thermal equilibriums,

Terminology

gradient
heat transfer
internal energy
thermodynamics
thermal equilibrium

and is therefore, by itself, of little help in determining quantitatively the transformations that occur from a lack of equilibrium in engineering processes. Since heat flow is the result of temperature nonequilibrium, its quantitative treatment must be based on other branches of science. The same reasoning applies to other types of transport processes such as mass transfer and diffusion.

Limitations of Classical

Classical thermodynamic deals with the states of systems from a macroscopic view and makes no hypotheses about the structure of matter. To perform a thermodynamics analysis it is necessary to describe the state of a system in terms of gross characteristics, such as pressure, volume, and temperature, that can be measured directly and involve no special assumptions regarding the structure of matter. These variables or thermodynamics properties are of significance for the system as a whole only when they are uniform throughout it, that is, when the system is in equilibrium. Thus, classical thermodynamic is not concerned with the details of a process but rather with equilibrium states and the relations among them. The processes employed in a thermodynamic analysis are idealized processes devised to give information concerning equilibrium states.

From a thermodynamic viewpoint, the amount of heat transferred during a process simply equals the difference between the energy change of the system and the work done. It is evident that this type of analysis considers neither the mechanism of heat flow nor the time required to transfer the heat. It simply prescribes how much heat to supply to or reject from a system during a process between specified end states without considering whether, or how, this could be accomplished. The fact that information cannot be obtained from a thermodynamic analysis considers neither the mechanism of heat flow nor the time required to transfer the heat. The question of how long it would take to transfer a specified amount of heat, although

Terminology

diffusion
macroscopic

it is of great practical importance, does not usually enter into the thermodynamic analysis.

Engineering Heat Transfer

From an engineering viewpoint, the key problem is the determination of the rate of heat transfer at a specified temperature difference. To estimate the cost, the feasibility, and the size of equipment necessary to transfer a specified amount of heat in a given time, a detailed heat transfer analysis must be made. The dimensions of boilers, heaters, refrigerators, and heat exchangers depend not only on the amount of heat to be transferred under given conditions. The successful operation of equipment components such as turbine blades or the walls of combustion chambers depends on the possibility of cooling certain metal parts by continuously removing heat from surface at a rapid rate. A heat transfer analysis must also be made in the design of electric machines, transformers, and bearings to avoid conditions that will cause overheating and damage the equipment. These examples show that almost every branch of engineering encounters heat transfer problems that are not capable of solution by thermodynamic reasoning alone but require an analysis based on the science of heat transfer.

In heat transfer, as in other branches of engineering, the successful solution of a problem requires assumptions and idealizations. It is almost impossible to describe physical phenomena exactly, and in order to express a problem in the form of an equation that can be solved it is necessary to make some approximations. In electrical circuit calculations, for example, it is usually assumed that the values of the resistances, capacitances, and inductances are independent of the current flowing through them. This assumption simplifies the analysis but may in certain cases severely limit the accuracy of the results.

Terminology

electric machine
heat exchanger
physical phenomena
capacitance
inductance
combustion chamber
transformer
resistance

It is important to keep in mind the assumptions, idealizations, and approximations made in the course of an analysis when the final results are interpreted. Sometimes insufficient information on physical properties makes it necessary to use engineering approximations to solve a problem. For example, in the design of machine parts for operation at elevated temperatures it may be necessary to estimate the proportional limit or the fatigue strength of the material from low-temperature data. To assure satisfactory operation of a particular part, the designer should apply a factor of safety to the results obtained from the analysis. Similar approximations are also necessary in heat transfer problems. Physical properties such as thermal conductivity or viscosity change with temperature, but if suitable average values are selected, the calculations can be considerably simplified without introducing an appreciable error in the final result. When heat is transferred from a fluid to a wall, as in a boiler, a scale forms under continued operation and reduces the rate of heat flow. To assure satisfactory operation over a long period of time, a factor of safety must be applied to provide for this contingency.

When it becomes necessary to make an assumption or approximation in the solution of a problem, the engineer must rely on ingenuity and past experience. There are no simple guides to new and unexplored problems, and an assumption valid for one problem may be misleading in another. Experience has shown, however, that the first requirement for making sound engineering assumptions or approximations is a complete and thorough physical understanding of the problem at hand. In the field of heat transfer, this means familiarity not only with the laws and physical mechanisms of heat flow but also with those of fluid mechanics, physics, and mathematics.

Heat transfer can be defined as the transmission of energy from one region to another as a result of a temperature difference between them. Since differences in temperatures exist all over the universe, the phenomena of

Terminology

viscosity
thermal conductivity
fluid mechanics
fatigue strength

heat flow are as universal as those associated with gravitational attractions. Unlike gravity, however, heat flow is governed not by a unique relationship but rather by a combination of various independent laws of physics.

The literature of heat transfer generally recognizes three distinct modes of heat transmission: conduction, radiation, and convection. Strictly speaking, only conduction and radiation should be classified as heat transfer processes, because only these two mechanisms depend for their operation on the mere existence of a temperature difference. The last of the three, convection, does not strictly comply with the definition of heat transfer because its operation also depends on mechanical mass transport. But since convection also accomplishes transmission of energy from regions of higher temperature to regions of lower temperature, the term"heat transfer by convection" has become generally accepted.

In the next three sections we will survey the basic equations governing each of the three modes of heat transfer. Our initial aim is to obtain a broad perspective of the field without becoming involved in details. We shall therefore consider only simple cases. Yet it should be emphasized that in most natural situations heat is transferred not by one but by several mechanisms operating simultaneously. Hence, we will show how to combine the simple relations in situations when several heat transfer modes occur simultaneously. Next we will illustrate how to use the laws of thermodynamics in heat transfer analysis, and in the last we will consider the units used in heat transfer calculations and methods for changing from one system of units to another.

Terminology

gravitational attraction
convection
heat flow
conduction
radiation

EXERCIS 2-2

A. Words and Phrases

1. take place	2. transit	3. deal with
4. neither ~ nor	5. quantitatively	6. restriction
7. feasibility	8. keep in mind	9. elevate
10. appreciable	11. be capable of	12. contingency
13. rely on	14. ingenuity	15. literature
16. transmission	17. simultaneously	18. at hand

B. Reading Comprehension

1. Put T or F to indicate if the sentences below are true or false according to the facts in the passage.

- Thermodynamics deals with the relation between heat and other forms of energy. ______________
- Heat cannot be measured directly. ______________
- Classical thermodynamics focuses on the microscopic structure. ______________
- A thermodynamic analysis considers either the mechanism of heat flow or the time required to transfer the heat. ______________
- There are three modes of heat transmission in the literature of heat transfer. ______________
- The principles of heat transfer can be derived from the basic laws of thermodynamics. ______________
- Heat flow is the result of temperature nonequilibrium. ______________

- The question of how long it would take to transfer a specified amount of heat does not usually enter into the thermodynamic analysis. ________________

2. Answer the following questions.

- What is the branch of science that deals with the relation between heat and other forms or energy?
- When does the heat transfer take place?
- Is it possible to measure the amount of heat transferred during a process by classical thermodynamics?
- List three kinds of heat transfer modes.

bibliography

Frank Kreith and Mark S. Bohn, *Principles of Heat Transfer*, pp. 1~4, Info Acess and Distribution Pte, 1993.

Unit

03

Fluid Flow and Basic Units

Unit 03 Fluid Flow and Basic Units

1) Some Characteristics of Fluids

One of the first questions we need do explore is, What is a fluid? Or we might ask, what is the difference between a solid and a fluid? We have a general, vague idea of the difference. A solid is "hard" and not easily deformed, whereas a fluid is "soft" and is easily deformed(we can readily move through air). Although quite descriptive, these casual observations of the differences between solids and fluids are not very satisfactory from a scientific or engineering point of view. A closer look at the molecular structure of materials reveals that matter that we commonly think of as a solid(steel, concrete, etc.) has densely spaced molecules with large intermolecular cohesive forces that allow the solid to maintain its shape, and not to be easily deformed. However, for matter that we normally think of as a liquid(water, oil, etc), the molecules are spaced farther apart, the intermolecular forces are smaller than for solids, and the molecules have more freedom of movement. Thus, liquids can be easily deformed(but not easily compressed) and can be poured into containers or forced through a tube. Gases(air, oxygen, etc) have even greater molecular spacing and freedom of motion with negligible cohesive intermolecular forces and as a consequence are easily deformed(and compressed) and will completely fill the volume of any container in which they are placed.

Terminology

solid
cohesive force
molecular structure
intermolecular
fluid

Although the differences between solids and fluids can be explained qualitatively on the basis of molecular structure, a more specific distinction is based on how they deform under the action of an external load. Specifically, a fluid is defined as a substance that deforms continuously when acted on by a shearing stress of any magnitude. A shearing stress(force per unit area) is created whenever a tangential force acts on a surface. When common solids such as steel or other metals are acted on by a shearing stress, they will initially deform (usually a very small deformation), but they will not continuously deform (flow). However, common fluids such as water, oil, and air satisfy the definition of a fluid — that is they flow when acted on by a shearing stress. Some materials, such as slurries, tar, putty, toothpaste, and so on, are not easily classified since they will behave as a solid if the applied shearing stress is small, but if the stress exceeds some critical value, the substance will flow. The study of such materials is called rheology and does not fall within the province of classical fluid mechanics. Thus, all the fluids we will be concerned with in this text will conform to the definition of a fluid given previously.

Terminology

fluid mechanics
rheology
tangential force
shearing stress
macroscopic

Although the molecular structure of fluids is important in distinguishing one fluid from another, it is not possible to study the behavior of individual molecules when trying to describe the average, or macroscopic, value of the quantity of interest, where the average is evaluated over a small volume containing a large number of molecules. Thus, when we say that the velocity at a certain point in a fluid is so much, we are really indicating the average velocity of the molecules in a small volume surrounding the point. The volume is small compared with the physical dimensions of the system of interest, but large compared with the average distance between molecules. Is this a reasonable way to describe the behavior of a fluid? The answer is generally yes, since the spacing between molecules is typically very small. For gases at normal pressures and temperatures, the spacing is on the order of 10^{-6} mm, and for liquids it is on the order of 10^{-7} mm. The number of molecules per cubic millimeter is on the order of 10^{18} for gases and 10^{21} for liquids. It is thus clear that the number of molecules in a very tiny volume is huge and the idea of using average values taken over this volume is certainly reasonable. We thus assume that all the fluid characteristics we are interested in (pressure, velocity, etc) vary continuously throughout the fluid — that is, we treat the fluid as a continuum. This concept will certainly be valid for all the circumstances considered in this text. One area of fluid mechanics for which the continuum concept breaks down is in the study of rarefied gases such as would be encountered at very high altitudes. In this case the spacing between air molecules can become large and the continuum concept is no longer acceptable.

Terminology

rarefied gas
continuum

EXERCIS 3-1

A. Words and Phrases

1. vague
2. descriptive
3. consequence
4. distinction
5. be concerned with
6. distinguish ~ from
7. continuum

B. Reading Comprehension

1. Put T or F to indicate if the sentences below are true or false according to the facts in the passage.

- The space of molecules in solids is farther apart than in fluids. ________
- Liquids can be easily deformed and can be poured into containers. ________
- In general, we can treat fluid as a continuum. ________
- If a substance flows when acted on by a shearing stress, it is a fluid. ________
- The rarefied gases can be encountered at low altitudes. ________
- A shearing stress is created whenever a tangential force acts on a surface. ________
- Such material as toothpaste, slurries and putty can be dealt with in classical fluid mechanics. ________
- The spacing between gas molecules at normal pressure is on the order of 10^{-6} mm. ________

2. Answer the following questions.

- Explain the differences between solids and fluids.
- What is the reason for using average values taken over tiny volume in fluid mechanics instead of using values of individual molecules ?
- Why does the continuum concept break down in the study of rarefied gases?
- What is the number of molecules per cubic millimeter in liquids?

bibliography

Bruce R. Munson, Donald F. Young, Theodore H. Okiishi, *Fundamentals of Fluid Mechanics*, pp. 2~3, John Wiley & Son, 2006.

2 Basic Units And Fluid Properties

Every fluid has certain characteristics by which its physical condition may be described. We call such characteristics properties of the fluid. These properties are expressed in terms of a limited number of basic dimensions (length, mass or force, time, and temperature), which in turn are quantified by basic units. The traditional system of units in the United States has been the foot-pound-second system. However, because all the engineering societies are urging the adoption of the SI(Systeme International) system, this text uses both SI and traditional units. Approximately half of the problems and the majority of the examples are given in the SI system; the remainder are given in the foot-pound-second system.

SI System of Units

Terminology

newton (N)
traditional units
SI
Newton's second law

The basic units of mass, length, and time in the SI system are the kilogram (kg), meter (m), and second (s). The corresponding unit of force is derived from Newton's second law: the force required to accelerate a kilogram at one meter per second is defined as the newton (N). The acceleration due to gravity varies by 0.5% over the earth's surface, and the standard accepted value is 9.80655 m/s^2 or 9.81 m/s^2 for engineering calculations. Thus the weight of a kilogram at the earth's surface is

$$\begin{aligned} W &= Mg \\ &= (1)(9.81)\, kg \cdot m/s^2 = 9.81\ \mathrm{N} \end{aligned} \tag{3.1}$$

From Eq. (3.1) we can also determine the units for kilograms in terms of the meter-newton-second units. From Eq. (3.1)

$$M = \frac{W}{g} \frac{N}{m/s^2}$$

Thus the mass M of a body in kilograms is given by its weight in newtons at the earth's surface divided by 9.81 m/s^2 (the acceleration due to gravity at the earth's surface). As shown, the units of mass may also be expressed in equivalent meter-newton-second units as $N \cdot s^2/m$.

The basic unit of temperature in the SI system is the Kelvin (K). The Kelvin scale is 0 K at absolute zero and 273.16 K at the freezing point of water. The Celsius scale (℃) is 0 ℃ at the freezing point of water. Therefore the conversion formula is

$$K = 273 + ℃$$

The unit of work and energy in the SI system is the *joule* (J), which is a newton meter ($N \cdot m$). The unit of power is the *watt* (W), which is a joule per second (J/s).

The expressions for power of 10 are

G (giga)	=	10^9	c (centi)	=	10^{-2}
M (mega)	=	10^6	m (milli)	=	10^{-3}
k (kilo)	=	10^3	μ (micro)	=	10^{-6}

For example, km stands for kilometer, or 1000 meters, and mm signifies millimeter, or 0.001 meter.

Traditional Units

The system of units that preceded SI in several countries is the so-called English system. The fundamental units of length and mass are the foot (ft), equal to 30.48 cm, and the slug, equal to 14.59 kg. The time unit of second

Terminology

slug
Celsius scale (℃)
gravity
watt (W)
English system
joule (J)
Kelvin (K)
freezing point

is common to both systems. The force required to accelerate a mass of one slug at one foot per second is one pound force (lbf). The mass unit more common to mechanical engineers in the traditional system is the pound mass (lbm). The conversion factor for changing pounds mass to slugs is g_c, which is equal to 32.2 lbm/slug. In other words,

$$1 \text{ slug} = 32.2 \text{ lbm}$$

In the traditional system it is good practice to use the units lbm and lbf and not simply lb.

The fundamental unit of work in the traditional system is the food-pound (ft-lbf), and the fundamental unit of power is the ft-lbf/s. A common unit of power is the horsepower, which is 550 ft-lbf/s. One horsepower is 746 W.

The traditional unit for temperature is the degree Fahrenheit (°F), which is $\frac{5}{9}$ of the Celsius degree. The corresponding absolute temperature scale is in degrees Rankine (°R). The Fahrenheit temperature at the freezing point of water is 32°F, and the formula for conversion between °F and °R is

Terminology

horsepower
Fahrenheit (°F)
Rankine (°R)

$$°R = 460 + °F$$

Properties Involving the Mass or Weight of the Fluid

Mass Density, ρ

The mass per unit volume is *mass density* and has units of kilograms per cubic meter or pounds mass per cubic foot. The mass density of water at 4℃ is 1000 kg/m^3 (62.4 lbm/ft^3) and decreases slightly with increasing temperature, as shown in Table A.5. The mass density of air at 20℃ and standard atmospheric pressure is 1.2 kg/m^3 (0.075 lbm/ft^3). Mass density, often simply called density, is represented by the Greek symbol ρ (rho). The densities of common fluids are given in Tables A.2 to A.5.

Specific Weight, γ

The gravitation force per unit volume of fluid, simply the weight per unit volume, is defined as specific weight. It is given the symbol γ (gamma). Water at 20℃ has a specific weight of 0.79 kN/m^3. In contrast, the specific weight of air at the same temperature and at standard atmospheric pressure is 11.8 N/m^3. Specific weight and density are related by

$$\gamma = \rho g$$

Specific weight of common liquids are given in Table A.4 in the Appendix.

Variation in Density

The density of some fluids is more easily changed than that of others. For example, air can be compressed at moderate pressure with a consequent density change, whereas a very large pressure is needed to effect a relatively small density change in water. For most applications, water can be considered in compressible and, in turn, can be assumed to have constant density. Air, on the other hand, is a relatively compressible fluid with variable density. However, at velocities much less than the speed of sound, the air density changes only slightly and the air can also be treated as incompressible.

Incompressibility does not always imply constant density. For example, a mixture of salt in water changes the density of the water without changing its volume. Therefore there are some flows, such as in estuaries, in which density is changing. Such a fluid is termed nonhomogeneous. This text emphasizes the flow of homogeneous fluids, so the term *incompressible*, used throughout the text, implies constant density.

Terminology

atmospheric
compressible
specific weight
incompressibility
gravitation force
nonhomogeneous

Specific Gravity, S

The ratio of the specific weight of a given fluid to the specific weight of water at a standard reference temperature is defined as specific gravity, S:

$$S = \frac{\gamma_{fluid}}{\gamma_{water}} = \frac{\rho_{fluid}}{\rho_{water}}$$

The standard reference temperature for water is often taken as 4℃, where the specific weight of water at atmospheric pressure is 9810 N/m^3. With this reference, the specific gravity of mercury at 20℃ is

$$S_{Hg} = \frac{133\,kN/m^3}{9.81\,kN/m^3} = 13.6$$

Because specific gravity is a ratio of specific weights, it is dimensionless and, of course, independent of the system of units used.

Terminology

thermal energy
specific gravity

Properties Involving the Flow of Heat

Specific Heat, c

The property that describes the capacity of a substance to store thermal energy is called specific heat. By definition, it is the amount of thermal energy that must be transferred to a unit mass of substance to raise its temperature by one degree.

The specific heat of a gas depends on the process accompanying the change in temperature. If the specific volume v of the gas ($v = 1/\rho$) remains constant while the temperature changes, then the specific heat is identified as c_v. However, if the pressure is held constant during the change in state, then the specific heat is identified as c_p. The ratio c_p/c_v is given the symbol k. Values for c_p and k for various gases are given in Table A.2.

Specific Internal Energy, u

The energy that a substance possesses because of the state of the molecular activity in the substance is termed internal energy. Internal energy is usually expressed as a specific quantity - that is, internal energy per unit mass. In the SI system, the specific internal energy, u, is given in joules per kilogram. The internal energy is generally a function of temperature and pressure. However, for an ideal gas, it is a function of temperature alone.

Specific Enthalpy, h

The combination $u + p/\rho$ is encountered frequently in the equations for thermodynamics and compressible flow; it has been given the name specific enthalpy. For an ideal gas, u and p/ρ are functions of temperature alone. Consequently their sum, specific enthalpy, is also a function solely of temperature.

Viscosity

The most important distinction between a solid such as steel and a viscous fluid such as water or air is that shear stress in a solid material is proportional to shear strain, and the material ceases to deform when equilibrium is reached, whereas the shear stress in a viscous fluid is proportional to the time rate of strain. The proportionality factor for the solid is the shear modulus. The proportionality factor for the viscous fluid is the dynamic, or absolute, viscosity. For example, near a wall, as shown in Fig 3.1, the rate of strain is dV/dy, where V is the fluid velocity and y is the distance measured from the wall, so the shear stress is

Terminology

shear modulus
joules
strain
enthalpy
shear stress
absolute
viscosity
internal energy

$$\tau = \mu \frac{dV}{dy}$$

where τ (tau) is the shear stress, μ (mu) is the dynamic viscosity, and dV/dy is the time rate of strain, which is also the velocity gradient normal to the wall. Thus the definition of the viscosity, μ, is the ratio of the shear stress to the velocity gradient, $\mu = \tau(dV/dy)$.

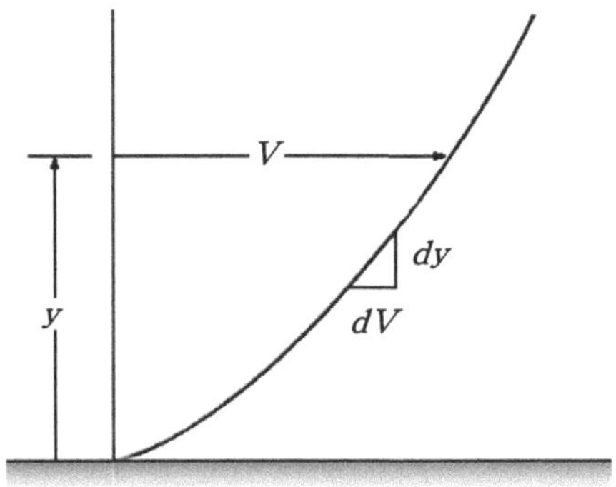

▲ **Figure 3.1** distribution next to a boundary.

Terminology

dynamic viscosity

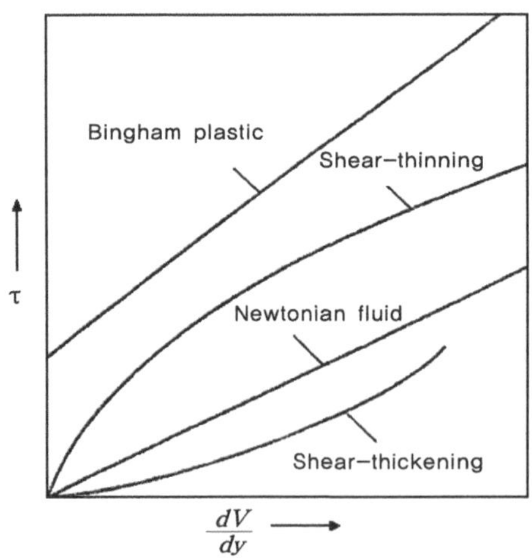

▲ **Figure 3.2** Shear–stress relations for different types of fluids.

Newtonian versus Non-Newtonian Fluids

Fluids for which the shear stress is directly proportional to the rate of strain are called Newtonian fluids. Because shear stress is directly proportional to the shear strain, dV/dy, a plot relating these variables (see Fig 3.2) results in a straight line passing through the origin. The slope of this line is the value of the dynamic viscosity. For some fluids the shear stress may not be directly proportional to the rate of strain; these are called non-Newtonian fluids. One class of non-Newtonian fluids, shear-thinning fluids, has the interesting property that the ratio of shear stress to shear strain decreases as the shear strain increases (see Fig 3.2). Some common shear-thinning fluids are toothpaste, catsup, paints, and printer's ink. Fluids for which the viscosity increases with shear rate are shear-thickening fluids. Some examples of these fluids are mixtures of glass particles in water and gypsum-water mixtures. Another type of non-Newtonian fluid, called a Bingham plastic, acts like a solid for small values of shear stress and then behaves as a fluid at higher shear stress. The shear stress versus shear strain rate for a Bingham plastic is also shown in Fig 3.4. This book will focus on the theory and applications involving Newtonian fluids. For more information on the theory of flow of non-Newtonian fluids, please see references (2) and (3).

Terminology

shear−thinning
Newtonian fluid
molecular attraction
Bingham plastic

Surface Tension

According to the theory of molecular attraction, molecules of liquid considerably below the surface act on each other by forces that are equal in all directions. However, molecules near the surface have a greater attraction for each other than they do for molecules below the surface. This produces a surface on the liquid that acts like a stretched membrane. Because of this membrane effect, each portion of the liquid surface exerts "tension" on adjacent portions of the surface or on objects that are in contact with the liquid surface. This tension acts in the plane of the surface, and its magnitude per unit length is defined as surface tension, σ (sigma). Surface tension for a water-air surface is 0.073 N/m at room temperature. The effect of surface tension is illustrated for the case of capillary action in a small tube (Fig 3.3). Here the end of a small-diameter tube is put into a reservoir of water, and the characteristic curved water surface occurs within the tube. The relatively great attraction of the water molecules for the glass cause the water surface to curve upward in the region of the glass wall. Then the surface-tension force acts around the circumference of the tube, in the direction indicated. It may be assumed that Θ (theta) is equal to 0° for water against glass. This produces a net upward force on the water that causes the water in the tube to rise above the water surface in the reservoir. An example gives a quantitative illustration of the principle.

Terminology

capillary action
surface tension

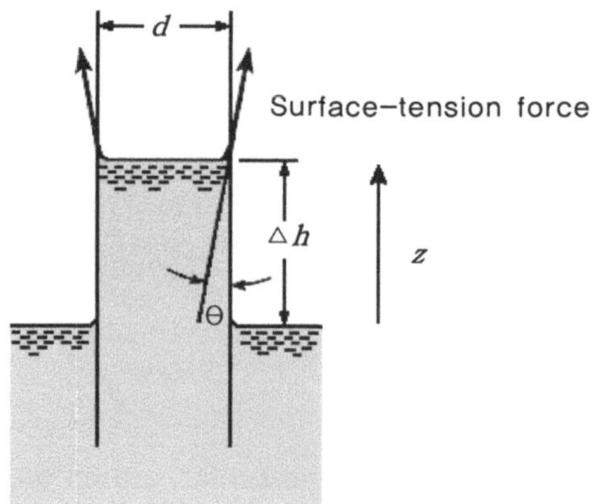

▲ Figure 3.3 Capillary action in a small tube.

Vapor Pressure

The pressure at which a liquid will boil is called its vapor pressure. This pressure is a function of temperature (vapor pressure increases with temperature). In this context we usually think about the temperature at which boiling occurs. For example, water boils at 212°F at sea-level atmospheric pressure (14.7 psia). However, in terms of vapor pressure, we can say that by increasing the temperature of water at sea level to 212°F, we increase the vapor pressure to the point at which it is equal to the atmospheric pressure (14.7 psia), so that boiling occurs. When we think of incipient boiling in terms of vapor pressure, it is easy to visualize that boiling can also occur in water at temperatures much below 212°F if the pressure in the water is reduced to its vapor pressure. For example, the vapor pressure of water at 50°F (10℃) is 0.178 psia (approximately 1% of standard atmospheric pressure). Therefore, if the pressure within water at that temperature is reduced to that value, the water boils.

- Such boiling often occurs in flowing liquids, such as on the suction side of a pump. When such boiling does occur in flowing liquids, vapor bubbles start growing in local regions of very low pressure and then collapse in regions of higher pressure downstream.
- Actually, boiling can occur at this vapor pressure only if there is a gas-liquid surface present to allow the process to start. Boiling at the bottom of a pot water is usually initiated in crevices in the material of the pot, in which minute bubbles of air are entrapped even when the pot is filled with water.

Terminology

vapor pressure

EXERCIS 3-2

A. Words and Phrases

1. adoption	2. equivalent	3. derivation
4. in contrast	5. entrap	6. accompany
7. gradient	8. be tangent to	9. be dependent on
10. be proportional to	11. adjacent	12. reservoir
13. circumference	14. incipient	15. collapse
16. downstream	17. crevice	18. membrane

B. Reading Comprehension

1. Put T or F to indicate if the sentences below are true or false according to the facts in the passage.

- The unit system currently used in the United States is the SI system. ____________
- The joule is the unit of work and energy in the SI system. ____________
- Slug is the force unit in English system. ____________
- The mass density of water increases slightly with increasing temperature. ____________
- At velocities much less than the speed of sound, the air can be treated as incompressible. ____________
- Specific heat describes the capacity of a substance to store thermal energy. ____________
- The English unit for specific internal energy is joules per kilogram. ____________

- The specific enthalpy is not a function solely of temperature.

  ____

- The proportionality factor for the viscous fluid is the shear modulus.

  ____

2. Answer the following questions.

- Explain the distinction of the proportionality factor between solids and liquids.
- What is the SI unit for specific weight?
- Convert 1 slug of mass in English system into the equivalent SI unit?
- List three basic units in the SI system.

bibliography

Clayton T. Crowe, Donald F. Elger, and John A. Roberson, *Engineering Fluid Mechanics*, pp. 10~26, John Wiley & Sons, 2001.

Unit 04

Engineering Materials

Unit 04 Engineering Materials

1 Characteristics of Engineering Materials

One of the simplest loading conditions of a material is that of *tension*, in which the fibres of the material are stretched. Consider, for example, a long steel wire held rigidly at its upper end, Fig 4.1, and loaded by a mass hung from the lower end. If vertical movements of lower end are observed during loading it will be found that the wire is stretched by a small, but measurable, amount from its original unloaded length. The material of the wire is composed of a large number of small crystals

Terminology

crystal
loading condition

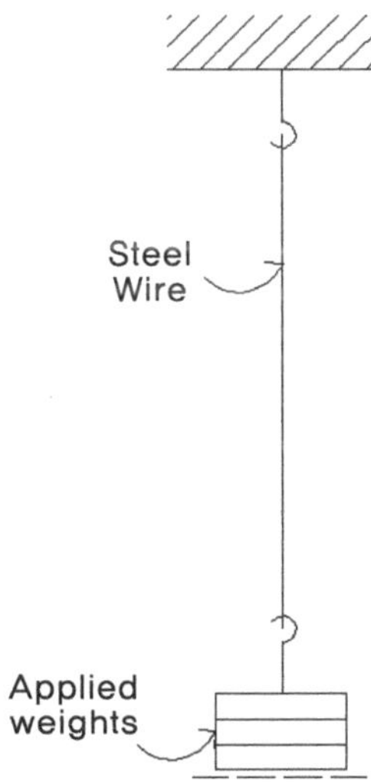

◀ **Figure 4.1** Stretching of a steel wire under end load

which are only visible under microscopic study; these crystals have irregularly shaped boundaries, and largely random orientations with respect to each other; as loads are applied to the wire, the crystal structure of the metal is distorted.

For small loads it is found that the extension of the wire is roughly proportional to the applied load, Fig 4.1. This linear relationship between load and extension was discovered by Robert Hooke in 1678; a material showing this characteristic is said to obey *Hooke's law*.

As the tensile load in the wire is increased, a stage is reached where the material ceases to show this linear characteristic; the corresponding point on the load-extension curve of Fig 4.1 is known as the limit of proportionality. If the wire is made from a high-strength steel then the load-extension curve up to the breaking point has the form shown in Fig 4.1. Beyond *the limit of proportionality* the extension of the wire increases non-linearly up to the elastic limit and, eventually, the *breaking point*.

The elastic limit is important because it divides the load-extension curve into two regions. For loads up to the elastic limit, the wire returns to its original unstretched length on removal of the loads; this property of a material to recover its original form on removal of the loads is known as *elasticity*; the steel wire behaves, in fact, as a stiff elastic spring. When loads are applied above the elastic limit, and are then removed, it is found that the wire recovers only part of its extension and is stretched permanently; in this condition the wire is said to have undergone an *inelastic, or plastic*, extension. For most materials, the limit of proportionality and the elastic limits are assumed to have the same value.

In the case of elastic extensions, work performed in stretching the wire is stored as *strain energy* in the material; this energy is recovered when the loads are removed. During inelastic extensions work is performed in making permanent changes in the internal structure of the material; not all the work performed during an inelastic extension is recoverable on removal

Terminology

load−extension curve
limit of proportionality
strain energy
elastic limit
elasticity
plastic
breaking point
microscopic study

of the loads; this energy reappears in other forms, mainly as heat.

The load-extension curve of Fig 4.2 is not typical of all materials; it is reasonably typical, however, of the behaviour of *brittle* materials.

An important feature of most engineering materials is that they behave elastically up to the limit of proportionality, that is, all extensions are recoverable for loads up to this limit. The concepts of linearity and elasticity form the basis of the theory of small deformations in stressed materials.

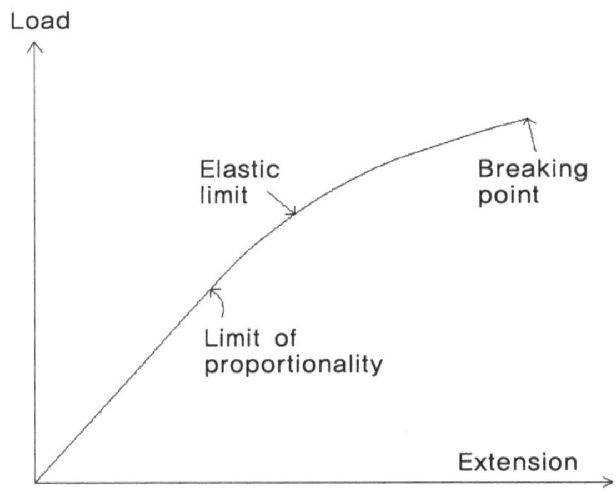

▲ **Figure 4.2** Load–extension curve for a steel wire showing the limit of linear–elastic behaviour (or limit of proportionality) and the breaking point

Terminology

brittle material
linearity
deformation

EXERCIS 4-1

A. Words and Phrases

1. be composed of
2. visible
3. random
4. with respect to
5. distort
6. cease to
7. be made from
8. divide ~ into
9. permanently
10. recoverable

B. Reading Comprehension

1. Put T or F to indicate if the sentences below are true or false according to the facts in the passage.

- Work is performed in making temporary changes in the internal structure of the material during inelastic extensions. ________________
- When a small tensile load is applied to a wire, the extension is roughly proportional to the applied load. ________________
- For most engineering materials, the material behave elastically up to the limit of proportionality. ________________
- When loads are applied above the elastic limit, the material undergoes a plastic deformation. ________________
- The material of the wire is composed of a small number of large crystals. ________________
- Hook's law can also be applied above the elastic limit. ________________

2. Answer the following questions.

- Explain the Hook's law in English.
- What is the elastic limit of a material? Explain in English.
- What happens to a wire under tensile strength above the elastic limit?

bibliography

John Case, Lord Chilver and Carl T. F. Ross, *Strength of Materials and Structures*, pp. 10~12, Edward Arnold, 1993.

2 Properties of Aluminum

Aluminum's position as the leading non-ferrous metal in volume of production is due largely to an impressive combination of qualities. These include strength with light weight, toughness, superior thermal conductivity, high electrical conductivity, high reflectivity, high corrosion resistance, excellent workability, ductility, superior finishing qualities, and pleasing appearance. Aluminum's nonmagnetic, and nontoxic properties, plus economy with availability extend its application to broad usage.

Strength with Light Weight

Light weight is perhaps aluminum's best known characteristic. With specific gravity of approximately 2.7, the metal weighs only about 0.1 pound per cubic inch as compared with 0.28 for iron and 0.32 for copper. Thus, in applications where volume remains the same regardless of metal used, aluminum goes about three times as far as either of these metals. Because of this low weight factor, combined with the high strength of many aluminum alloys, it is possible in many instances to substantially reduce the total weight of products and assemblies previously made from other metals.

Commercially pure aluminum, 1100, has a tensile strength of 13,000 psi. Its usefulness as a structural material in this form is somewhat limited. However, by cold rolling, aluminum's strength can be approximately doubled. Also, strength increases are obtained by adding alloying metals such as manganese, silicon, copper, magnesium, and zinc. Further, many of the aluminum alloys are strengthened and hardened by heat treatment. Some heat-treatable alloys obtain tensile strengths approaching 100,000 psi as the result of such treatment.

A wide variety of mechanical characteristics, or tempers, is available in aluminum alloys by subjecting the metal to both cold work and heat treatment.

Terminology

strengthen
harden
heat treatment
cold work
ductility
cold rolling
reflectivity
toughness
specific gravity
non−ferrous
thermal conductivity

Some aluminum alloys work harden more rapidly than others. Thus, in specifying the temper for any given application, the fabricating process and the amount of cold work to which the metal is to be subjected should be considered. The temper specified may be determined by the anticipated amount of cold work the metal receives during fabrication. When this is done, the desired formability and strength characteristics in the finished products will result.

Recent developments in alloys have also increased their dent resistance. Whereas the common conception of dent resistance of aluminum is often limited to that of cooking utensils or other products made from the softer alloys, modern high strength aluminum's dent resistance is often greater than many of the commonly used steels. Generally the dent resistance of an alloy is in direct proportion to its yield strength. Aluminum alloys that are fully heat treated have the best resistance to denting, but there are some of the common or non-heat-treatable alloys which, in their harder tempers, compare very favorably with some steels.

At sub-zero temperatures, aluminum's strength increases without loss of ductility so that it is a particularly useful metal for low temperature applications. The properties of aluminum alloys are not adversely affected by low temperatures. At temperatures down to -420°F, the properties of all alloys are at least equal to their room temperature properties and in most cases, are actually improved. No embrittlement of alloys at low temperatures is evidenced. The wrought alloys, in general, show improved elongation at low temperatures while most of the cast alloys show either a slight increase in elongation or no appreciable change. Notch sensitivity, as measured by the usual type of impact tests, is not adversely affected by low temperatures, and the modulus of elasticity, hardness, and fatigue strength are improved at lower temperatures.

Terminology

fatigue strength
embrittlement
dent resistance
modulus of elasticity
wrought alloy
cast alloy
elongation
hardness

Aluminum loses part of its strength at elevated temperatures, although numerous alloys retain good strength at temperatures from 400 to 500°F.

Applications should always be designed on the basis of room temperature properties.

Lightness with strength has made aluminum the basic material in aircraft production, is making it dominant in commercial highway transportation equipment, and is spurring its use for structural purposes. Lightness is also establishing this metal in many other important applications where the favorable strength weight ratio contributes to product improvement, better performance, and greater economy. Light weight, in many cases, adds to consumer acceptance of aluminum products through reduced handling and fabrication costs.

Thermal Conductivity

The characteristic high thermal conductivity of aluminum is important wherever the transfer of thermal energy from one medium to another is involved — whether in heating or cooling. On a weight basis, aluminum is the most efficient heat conductor of the common metals. This is one of the principal reasons it is so widely used for cooking utensils. Heat is rapidly and evenly conducted to the food being cooked without troublesome hot spots which may cause uneven cooking or scorching. Aluminum's lightness, workability, appearance, nontoxicity, an economy are also determining factors for this type of application.

Terminology

nontoxicity
condenser
evaporator
heat conductor

This metal is frequently used for engine cylinder heads because its high thermal conductivity makes greater compression ratios possible. Conductivity also influences aluminum's wide use in various parts and components such as ductwork, refrigerator evaporators, condenser coils, and similar equipment.

Other applications where aluminum's conductivity is utilized are in heat exchangers for the food, chemical, petroleum, air craft, and like industries.

Electrical Conductivity

Aluminum is the leading electrical conducting metal on a weight basis. Conventionally, electrical conductivity of a conductor material is expressed as a percentage of the International Annealed Copper Standard (IACS) on either a volume or a weight basis. Numerical value of this reference standard is 100 per cent at 20℃.

Annealed EC (electrical conductor) grade aluminum has an electrical conductivity at 20℃ of approximately 62 per cent on a volume basis. On a weight basis, it has some 200 percent conductivity.

This means that a section of aluminum of equal length and weight with copper has some 200 per cent conductance. It follows that an aluminum conductor of the same length and same conductance as a copper conductor weighs only one-half as much. It is this combination of high conductivity and low weight which has established aluminum so firmly in the transmission and distribution of electrical energy.

For certain applications such as long-span bus bars, mechanical rigidity is an important consideration. Certain high strength alloys are used for such applications in preference to EC, despite the fact that their conductivity is lower(53 to 57 percent on the volume basis, depending on the alloy selected).

Terminology

mechanical rigidity

EXERCISE 4-2

A. Words and Phrases

1. corrosion
2. workability
3. temper
4. conception
5. in proportion to
6. elongation
7. fabrication
8. utensil
9. petroleum
10. spur
11. conductor
12. formability
13. favorable
14. appreciable
15. contribute
16. in preference to

B. Reading Comprehension

1. Put T or F to indicate if the sentences below are true or false according to the facts in the passage.

- The tensile strength of pure aluminium can be doubled by cold working. ________________
- Aluminum's strength at sub-zero temperatures decreases. ________________
- Aluminum is frequently used for engine cylinder heads because of its high thermal conductivity. ________________
- Copper is the leading electrical conducting metal on a weight basis. ________________
- Notch sensitivity is adversely affected by high temperatures in aluminum. ________________
- Dent resistance of aluminum can be improved by heat treatment. ________________
- Some aluminum alloys obtain tensile strengths approaching 13,000 psi as a result of heat treatment. ________________

- Aluminum's best known characteristic is it's high thermal cond– uctivity.

  ____

- Copper is more frequently used in cooking utensils than aluminum.

  ____

2. Answer the following questions.

- What is one of the principal reasons aluminum is so widely used for cooking utensils?
- What makes aluminum the basic material in aircraft production?
- What are the advantages of using aluminum? List them all.
- List three properties of aluminum which are improved at low temperatures.
- How can the strength of aluminum increase?

bibliography

J.M. Adle, *Sheet and Plate*, pp. 1–11, Kaiser Aluminum & Chemical Sales, Inc 1953.

Unit 05

Fluid Power

Unit 05 Fluid Power

1) A Brief History of Fluid Power

Even before the written history of mankind, the power of air and water were put to use in an effort to make life and labor a little easier. The currents of flowing rivers and the wind were probably the first power sources harnessed by early man, primarily for transportation.

The first machines to utilize fluid power were water mills. These very large hydraulic motors used flowing water to turn the rotor (the water wheel). This rotated the shaft, which used gears, belts, pulleys, etc, to rotate the millstone to grind grains. Recorded history indicates that water mills were used as early as 100 BC, so they were probably around much earlier than that.

Hero of Alexandria was a scientist of the first century AD who is credited with the development of a number of early machines, including a reaction jet steam engine called an *aeolipile*. Steam was heated in the large container and piped into the sphere. The steam escaping through the two nozzles caused the sphere to rotate. This device was not used to do work. It was simply a novelty enjoyed by the nobility.

Another of Hero's inventions was a clever device for opening the doors of the temple. Air was heated in a closed container. The expansion of the air forced water through tubes to suspended containers that were attached by ropes and pulleys to the doors. As the containers filled, their weight

Terminology

belt
hydraulic motor
shaft
pulley
water mill
rotor
gear

slowly, quietly, and mysteriously opened the doors.

Hero is only one of the early scientists to make contributions to what we now call *fluid power*. Archimedes developed the screw pump in the third century BC. Greeks and Romans used water power in aqueducts and piping systems in which lead pipes carried water for both power and domestic use. The water clock — invented by Ktesbios around 250 AD — provided the most accurate method of keeping time until the Swiss movements were invented in the fourteenth century.

During the so-called Dark Ages, little technical progress of any kind occurred. Fluid power stagnated, as did all other scientific fields. The seventeenth century saw a new surge of scientific endeavors, and fluid power began to be seen as a major field of study. During that century, Sir Isaac Newton developed the concepts of *viscosity* and *drag*, along with his *Laws of Motion*. Torricelli did his work in fluid motion about the same time.

In the eighteenth century, Daniel Bernoulli, a Swiss scientist and scholar, studied energy in fluid streams and pipelines and developed the mathematical expressions for the transmission of fluid energy. A French scientist and mathematician, Blaise Pascal, developed the theories of hydrostatic pressure and force known as *Pascal's Law* (which we'll examine later). An Englishman, Joseph Bramah, used the principals of Pascal's Law and the related concepts of multiplications of force to develop the first hydraulic presses as the eighteenth century drew to a close.

Terminology

viscosity
treadmill
drag
Laws of Motion
water clock
hydrostatic pressure

The Industrial Revolution began in Britain in 1850. Interest in the use of fluid power to operate machines of all types expanded rapidly during this period. The forerunners of many of today's important hydraulically operated manufacturing tools were developed during that time. This was accompanied by the development of many different types of pumps—driven by hand- or foot-operated mechanisms, horse-operated turnstiles of treadmills, water wheels, windmills, steam engines, and other such devices—to provide hydraulic power for the machines.

One of the most remarkable developments was the pressurized water main system in the city of London. This system, which originated in 1848, was intended initially to supply hydraulic power to operate the dock elevators along the Thames River to ease the problems of moving cargo from the ships to the docks and then into upper floors of the warehouses. Initial progress was slow, but by 1883 there were 12 miles of pipeline laid out along the Thames supplying pressurized water at 700 psi. By 1920 this system had grown to 160 miles of piping providing hydraulic power for over 5000 freight elevators (with capacities to more than 40 tons), cranes and capstans, 1000 industrial presses, and 1500 passenger elevators. In addition, many thousands of small service elevators were in use by London merchants. The Great Stage at the Theatre Royal in London was powered from this main for many years.

▲ Figure 5.1 Hero's aeolipile was a reaction jet steam engine designed for entertainment, not work.
Source : Aircraft Gas Turbine Technology, 2nd ed., 1979, by I. Treager. Reprinted by permission of Glencoe Division of the Macmillan/McGraw-Hill School Publishing Company.

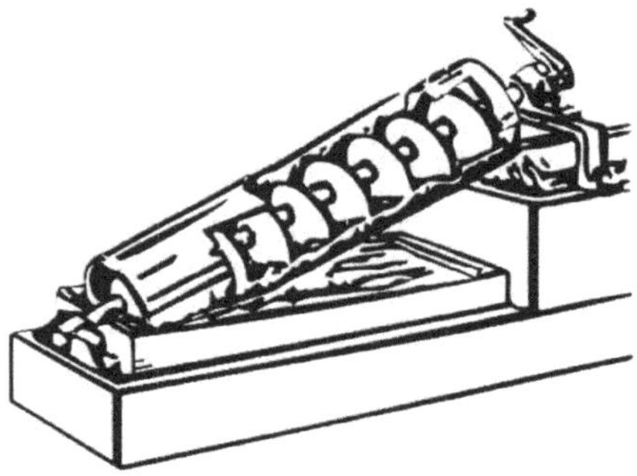

▲ **Figure 5.2** The Archimedean screw pump was invented in third century BC.
Source : From The American Peoples Encyclopedia, 1976 Edition, Copyright 1976, Permission of Grolier Incorporated.

Terminology

pneumatics
pneumatic fluid power
sewer system
rock drill

It was also during the nineteenth century that Navier expanded on Bernoulli's work by developing the equations for fluid flow that included the effects of friction. In the latter part of the century, Sir George Stokes, an Irish physicist and mathematician, independently confirmed Navier's work and continued to explore related topics. Osborne Reynolds, an English physicist, presented his work on the resistance to fluid flow and his definitions of laminar and turbulent flow late in the century.

The concept of *pneumatic fluid power*—the use of compressed air to do work—also blossomed in the nineteenth century. Early compressors had a maximum pressure of about 90 psi. One of the most important early applications of pneumatics was its use in the construction of an 8½ mile tunnel through Mt. Cenis in the French Alps. The project took 14 years to complete, which was less than half the original estimate based on the use of manual rock drills instead of pneumatic drills.

While the city of London was progressing with its pressurized water mains, the city of Paris was looking to compressed air mains. In 1888, 4 miles of mains with 30 miles of branch lines were established in an old sewer system. The 90 psi air was provided by a 65 HP compressor. In three years, the demand was so great that the compressors required over 25,000 HP. As with London's water mains, the compressed air service found many applications in both commercial and private sectors. Both the hydraulic and pneumatic mains gave way to electricity early in the twentieth century.

Fluid power became more sophisticated as the early twentieth century brought new materials and concepts. Water was replaced by petroleum-based oils with their improved viscosity and lubricity and their improved low-temperature characteristics. Higher pressures were possible because of improved sealing capabilities. Better machining processes provided closer tolerances and smaller clearances to reduce internal leakage, again allowing the use of higher pressures.

Strangely, while electricity was displacing pressurized water and compressed air as power distribution utilities, hydraulics and pneumatics were displacing electricity for heavy industrial and mobile applications. In 1906 the electric gun turret drives on the battleship USS *virginia* were replaced with hydraulic drives. The results were sufficiently impressive that virtually all subsequent ships' systems — propeller pitch control, steering systems, elevators, winches, etc — were (and still are) hydraulic.

Terminology

tolerance
clearance
water main
leakage
lubricity
steering system
winch

2 Fluid Power Applications

We are reminded of our dependence on electrical power. Just recall the inconvenience of the last brief power outage, with no light, no air conditioning(in summer), factories and businesses – or at least their computers – shut down. Few people, however, realize that fluid power plays just as important a role in our lives. Other than the brakes, automatic transmissions, and power steering on our cars, hydraulic jacks, and occasionally an elevator, few of us ever come into contact with hydraulic systems in our everyday home life. But there are very few segments of industry that do not depend to some extent on fluid power; and, of course, industry influences the daily lives of us all. Some applications are in industries where we might expect to find them–foundries, rolling mills, large cranes, bulldozers, etc. Other

Terminology

steering
transmission

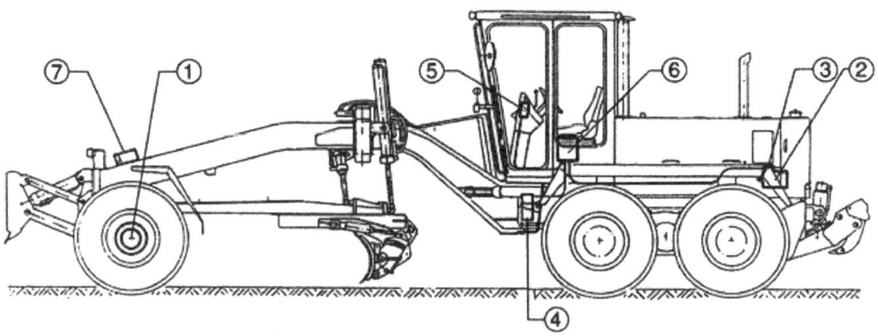

▲ **Figure 5.3** A modern road grader has several hydraulic components. ① Hydraulic wheel motors, ② variable displacement pump, ③ electronic proportional control valve, ④ solenoid valve bank, ⑤ control panel, ⑥ electronics, ⑦ flow divider.
source : The Rexroth Corporation

▲ **Figure 5.4** This large CNC lathe is hydraulically operated
Source : The Rexroth Corporation

applications are less well known—in textiles, food processing, automated monsters at theme parks, glass manufacturing, plastic fabricating, medical examination tables, garbage trucks, oil drilling rigs—the list goes on and on. Barbers' and dentists' chairs, for example, are raised and lowered by fluid power. In fact, almost 90 percent of American industry uses fluid power in some way. Figures 1.5 through 5.1 show some typical fluid power applications.

Terminology

drilling rig

EXERCIS 5-1,

A. Words and Phrases

1. in an effort to	2. harness	3. water mill
4. millstone	5. hydraulic	6. nobility
7. surge	8. viscosity	9. turnstile
10. capstan	11. pneumatic	12. sewer
13. tolerance	14. leakage	15. remind.... of ~
16. foundry	17. treadmill	18. give way to ~
19. novelty	20. draw to a close	21. turret
22. fabricate	23. outage	24. to some extent

B. Reading Comprehension

1. Put T or F to indicate if the sentences below are true or false according to the facts in the passage.

- Water mills were the first machines utilizing fluid power. ________
- Many people realize that fluid power plays an important role in our lives. ________
- The concept of pneumatic fluid power is to use compressed air in working. ________
- The city of London was progressing with its compressed air mains. ________
- Because of improved sealing capabilities, high pressure was possible to obtain. ________
- The base of hydraulic power is Pascal's law. ________

- By 1883, there were 12 miles of pipeline laid out along the Thames.
  ____
- Daniel Bernoulli first invented a reaction jet steam engine.
  ____

2. Answer the following questions.

- Who was the man expanded on Bernoulli's work by developing the equations for viscous fluid flow?
- List up some fluid power application In industries.
- Explain the difference in using fluid power between the city of London and the city of Paris.

3 The Fluid Power Industry

The fluid power industry can be divided into six broad categories:

- Fluid manufacturers
- Component manufacturers
- Original equipment manufactures (OEMs)
- Distributors
- Users of fluid power-driven machines
- Component repair facilities

The first three categories represent the "hard core" of the industry. It is within this group that most of the trained fluid power engineers and technologists are currently found. According to recent information from the National Fluid Power Association, this group consists of approximately 800 companies with 1200 plants employing some 340,000 people in the United States. The industry accounts for over $8 billions in sales annually.

Terminology

distributor
mechanic
pipefitter

Fluid power distributors provide the interface between those who manufacture fluid power products and those who use the products. While many people consider the distributors to be hardware salesmen, they are actually very talented and competent systems engineers and designers. They provide assistance in the design of fluid power systems as well as the modification and modernization of existing equipment. While they do sell and service components, the demands of their customers require that a large percentage of their time be spent in the engineering design function. Most fluid power distributors are members of the Fluid Power Distributors Association (FPDA).

The users group includes a major proportion of industry, but it actually employs a relatively small number of fluid power engineers and technologist.

Rather, it employs a large number of mechanics and pipefitters to maintain its fluid power systems. These maintainers are often relatively untrained in fluid power, which poses huge problems in keeping today's sophisticated, high pressure, often electronically controlled systems operating properly.

The last category — component repair facilities — can provide a invaluable service to the user group. In many cases, hydraulic component that have been removed from service can be repaired and put back into services for a fraction of the cost of a new component. It is often not practical, however, for users to maintain their own repair facilities, so they may choose to use the services of a *rebuild house* for overhaul. Many of these exist, offering factory-trained technicians, state-of-the-art test equipment, and shop facilities to properly service and rebuild components at a significant saving for the user.

Terminology

rebuild house
overhaul
repair facility

4 Future of Fluid Power

Recessions have changed the face of the fluid power industry. Many small companies have closed their doors permanently because of these business cycle. Others have merged or been acquired by larger, more stable companies. Even some of the giants of the industry have been forced to reduce their staffs and eliminate marginal product lines. The progressive companies, however, have also taken advantage of the lull in activity to

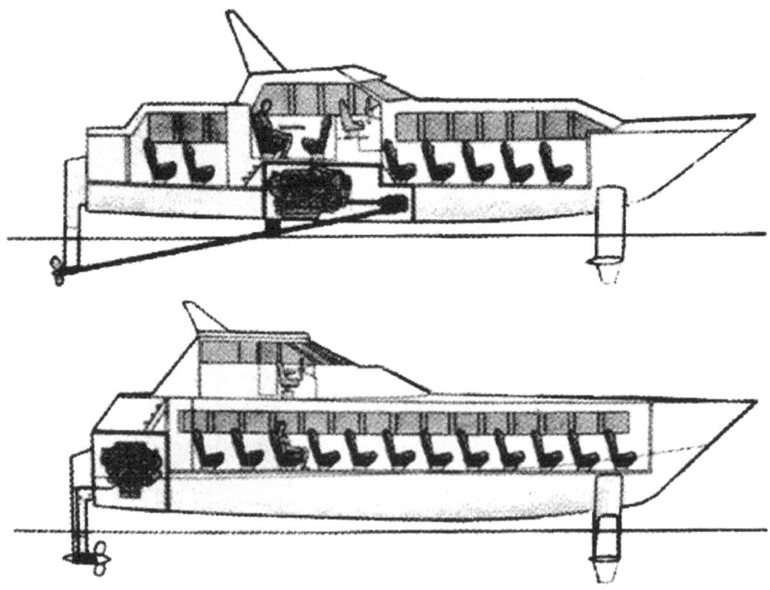

▲ Figure 5.5 The use of a hydraulic–motor–driven propellor on this modern hydrofoil (bottom) gives a 25% increase in carrying capacity and a 9% increase in speed over the traditional configuration (top). Both boats use the same engine.
Source : The Rexroth Corporation

regroup, reevaluate their futures, and even do some serious research and development that earlier rush to keep up with production had not allowed. The result has been a technologically stronger industry with more progressive leadership and a better vision of what will be required to compete with other modes of machine power(primarily electricity) and with international competition. For example, significant research has been done in the marriage of microprocessor controls and fluid power. Herein lies the real future of fluid power. The precision of electronic control applied to the mechanical speed and power of hydraulics and pneumatics will provide a versatility never before available to industry.

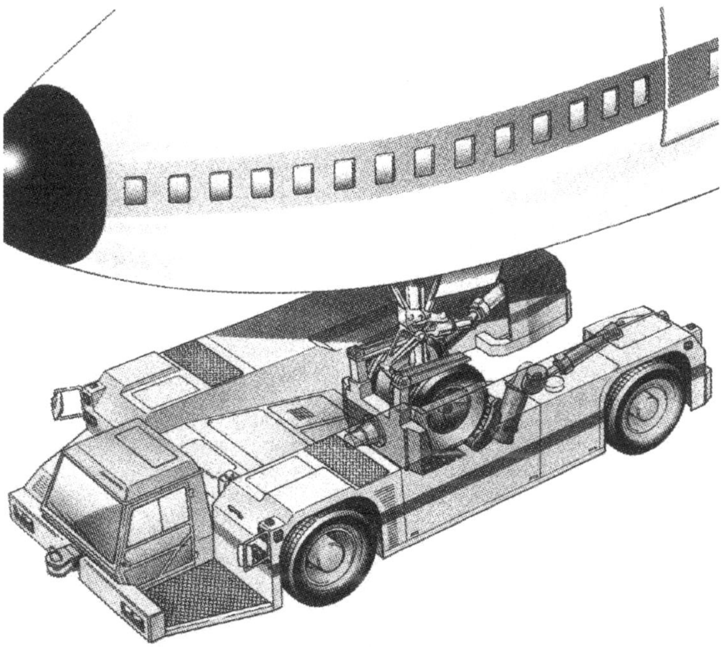

▲ **Figure 5.6** A ground mover system for large aircraft use hydraulic cylinders to lift and clamp the aircraft nose wheel. Hydraulic motors drive.
Source : The Rexroth Corporation

The prospects are exciting! Imagine huge manufacturing robots able to place and hold large loads weighing hundreds of pounds with a precision measured in thousandths of an inch. Consider programming an excavator to cut a trench to an exact depth and along a path that requires numerous changes in direction and do it all automatically. Lasers have already been used to control a motor grader to ensure precision leveling of an area.

The future of fluid power belongs to those engineers and technologists who are able to look past the traditional ways of designing systems and imagine new approaches to control, materials, reliability, energy conservation, filtration, health monitoring, maintainability, and all other such aspects. The future of fluid power depends on the ability — and inclination — of the people in the industry to stay on the leading edge of technology.

Terminology

maintainability
energy conservation
health monitoring
filtration
reliability
grader

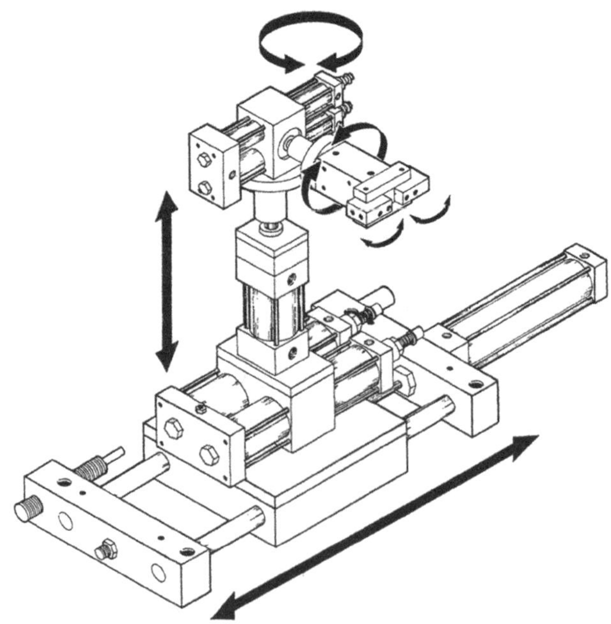

▲ Figure 5.7 This multi–axis, multi–rotation pick–and–place mechanism uses fluid power.
Source : PHD Inc. Fort Wayne, IN.

EXERCIS 5-3,

A. Words and Phrases

1. category
2. distributor
3. merge
4. marginal
5. versatility
6. trench
7. excavator
8. filtration
9. maintainability
10. modification
11. sophisticated
12. recession
13. take advantage of
14. lull
15. keep up with
16. compete with
17. ensure

B. Reading Comprehension

1. Put T or F to indicate if the sentences below are true or false according to the facts in the passage.

- The role of the fluid power equipment distributors is limited to the sales of hardware. ________
- Many small fluid power companies have been acquired by larger companies because of recession. ________
- Hydraulic component removed from service can be repaired by component repair facilities. ________
- Users of the equipment sometimes had better choose to use the services of a rebuild house for overhaul. ________
- The users group always maintain relatively well trained technician. ________

2. Answer the following questions.

- What are the six broad categories of the fluid power industry?
- What is the role of fluid power distributors?
- What does the future of fluid power depend on?
- Which categories of the fluid power industry are called "hard core"?

bibliography

F.D Norvelle, *Fluid Power Technology*, pp. 1–4, west, 1995.

Unit

06

Energy

Unit 06 Energy

1) Energy : An Initial Definition

Energy is one of the major building blocks of modern society. Energy is needed to create goods from natural resources and to provide many of the services we have come to take for granted. Economic development and improved standards of living are complex processes that share a common denominator: the availability of an adequate and reliable supply of energy. The modernization of the West from a rural society to an affluent urban one was made possible through the employment of modern technology based on a multitude of scientific advances—all of which are energized by fossil fuels. Political events, beginning with an oil embargo in 1973 and continuing through the Iranian revolution of 1979 and the Persian Gulf War of 1991, made many people aware of how crucial energy is to the everyday functioning of our society. Long gasoline lines and cold winters with natural gas shortages in the 1970s are still unhappy memories for some people. The energy crises of the 1970s were almost forgotten by the 1980s. However, that decade brought an increased awareness of our environment. Concerns about global warming, acid rain, and radioactive waste are still very much with us today, and each of theses topics is related to our use of energy.

Terminology

denominator
fossil fuel
global warming
acid rain
radioactive waste

Energy supplies are key limiting factors to economic growth. We have become a very interdependent world, and access to adequate and reliable energy resources is central for economic growth. About 40% of the world's energy comes from oil much of which is imported by the industrialized nations and much of which comes from the Persian Gulf. From this region, Japan imports two thirds of its oil, the United States imports 20% of its oil, while one third of France's total oil needs comes from there. If industrialized nations encounter any significant restriction to their sources of oil, through either reduced supplies or large price increases, their economies would suffer considerable damage.

Your own picture of energy might be colored in many ways by your experiences. You might think of the "energy" (or the lack of it) that a particular person possesses, or the kinetic energy that a stone gains as it drops, or the energy responsible for the movement of automobiles, or the energy used in the production of heat and light. One dictionary defines energy as the "capacity for vigorous action; inherent power; potential forces." Energy is found in many forms, and one purpose of this book will be to identify them and study how they can be used. Energy is found in such forms as wind and flowing water, and stored in matter such as fossil fuels—oil, coal, natural gas—where it can be burned for "vigorous action."

Energy might best be described in terms of what it can do. We cannot "see" energy, only its effects; we cannot make it, only use it; and we cannot destroy it, only waste it (that is, use it inefficiently). Unlike food and housing, energy is not valued in itself but for what can be done with it.

Energy is an end in itself (notes Richard Balzhiser, former president of Electric Power Research Institute). The fundamental goals we should have in mind are a healthy economy and a healthy environment. We have to tailor our energy policy as a means to those ends, not just for this country but in global terms as well.

Energy is a basic concept in all the sciences and engineering disciplines. As we will discuss in the next chapter, a very important principle is that energy is a conserved quantity, that is, the total amount of energy in the universe is a constant. Energy is not created or destroyed but just converted or redistributed from one form to another, such as from wind energy into electrical energy, or from chemical energy into heat. We will study the various forms of energy—chemical, nuclear, solar, thermal, mechanical, electrical—and the useful work that energy is capable of doing for us. We will explore both energy resources and energy conversion processes.

Understanding energy means understanding energy resource and their limitations, as well as the environmental consequences of their use. Energy and environment and economic development are closely linked. Over the past two decades, global energy consumption has increased by about 25%, while U.S. consumption increased by 15%. Much of this global growth has been in less developed countries. (In the next two decades, estimates are that energy consumption will rise by over 100% in developing nations.) With this growth has been a decline in urban air quality as well as serious land and water degradation. Since fossil fuels represent almost 90% of our consumption, we continue to increase the emissions of carbon dioxide, which may alter the earth's climate irreversibly. The proper use of energy requires consideration of social issues as well as technological ones. Indeed, sustained economic growth in this century, together with improvements in the quality of everyone's lives, may be possible only by the well-planned and efficient use of limited energy resources.

Terminology

carbon dioxide
irreversibly
energy resource

EXERCIS 6-1

A. Words and Phrases

1. take for granted	2. share	3. urban
4. embargo	5. affluent	6. awareness
7. shortage	8. radioactive	9. lack of
10. vigorous	11. have in mind	12. discipline
13. degradation	14. irreversibly	15. sustain
16. standard of living	17. availability	18. rural
19. crucial	20. suffer	

B. Reading Comprehension

1. Put T or F to indicate if the sentences below are true or false according to the facts in the passage.

- Energy is an end in itself. ____________
- Much of the global growth of energy consumption has been in developed countries such as United States. ____________
- Energy is not always conserved in classical physics. ____________
- Economic growth cannot be limited by energy supplies. ____________
- The modernization of West was possible through the employment of modern technology energized by fossil fuels. ____________
- Carbon dioxide may alter the earth's climate. ____________
- The proper use of energy requires only consideration of technological issues. ____________
- A common denominator of economic development and improved standards of living is the availability of an adequate and reliable supply of energy.

____

- About 50% of the world's energy comes from oil. ________________

2. Answer the following questions.

- By how much percentage the energy consumption will rise in developing nations in the next two decades?
- How long had the oil embargo in 1973 last?
- What are our major concern related to our use of energy? List three.
- List some major countries that depend on the Persian Gulf in much of their energy consumption.
- How is the term 'energy' defined in the dictionary?

2 Work, Energy, and Machines

If force is considered a measure of effort, then work is a measure of accomplishment and energy is the ability to do work. The concept of work makes no allowance for a time factor. When a time factor is considered, the concept of power must be introduced. Power is work per unit time.

In the following pages the concepts of work, energy, and power(which are normally presented in college physics course) are covered in some detail, after which a method, based on the law of conservation of energy, for obtaining equations of motion of systems is presented.

Work

The work done in a mechanical system is the product of force multiplied by the distance(or torque multiplied by the angular displacement) through which it is exerted, both force and distance being measured in the same direction. For instance, if a body is pushed with a horizontal force of F newtons along a horizontal floor for a distance of x meters, the work done W in pushing the body is

$$W = F \cdot x \ (Nm)$$

Terminology

equations of motion
power
horizontal force
work
energy
conservation of energy
torque
angular displacement

Energy

In a general way, energy can be defined as the capacity or ability to do work. It is found in many different forms and can be converted from one form into another. For instance, an electric motor converts electrical energy into mechanical energy, a battery converts chemical energy into electrical energy, and so forth.

A system is said to possess energy when it can work. When a system does mechanical work, the system's energy decreases by the amount equal to the energy required for the work done. Units of energy are the same as units for work, that is, newton-meter, joule, kcal, Btu, and so on.

According to the law of conservation of energy, energy can be neither created nor destroyed. This means that the increase in the total energy within a system is equal to the net energy input to the system. So if there is no energy input, there is no change in total energy of the system.

The energy that a body possesses because of its position is called potential energy, whereas the energy that a body has as a result of its velocity is called kinetic energy.

A machine is a device that helps make work easier to perform. (Remember, work = force x distance). A machine makes work easier to perform by accomplishing one or more of the following functions:

transferring a force from one place to another,
changing the direction of a force,
increasing the magnitude of a force, or
increasing the distance or speed of a force

Terminology

potential energy
kinetic energy
mechanical advantage

Mechanical Advantage

When a machine takes a small input force and increases the magnitude of the output force, a mechanical advantage has been produced. If a machine increases an input force of 10 pounds to an output force 100 pounds, the machine has a mechanical advantage (MA) of 10.

Conservation of Energy

No machine can increase both the magnitude and the distance of a force at the same time. When a machine produces an increase in force, there is always a proportional decrease in the distance moved. Conversely, when a machine produces an increase in distance, there will be a proportional decrease in force.

Another way to state this concept is that no machine can produce more work than the amount of work that is put into the machine. In fact, if you ignore the work lost due to friction and its inefficiencies, the amount of work produced by a machine is always exactly the same as the amount of work put into the machine. This is known as the law of conservation of energy.

This concept can also be put into the form of an equation. (Remember that work is equal to force times distance.)

$$F1 \times D1 = F2 \times D2$$

where, F1 = Input Force F2 = Output Force,
D1 = Input Distance D2 = Output Distance.

Terminology

wedge
complex machine
simple machine
screw
axle
friction
pulley

The Simple Machines

You are probably familiar with many different machines. Some of thesc machines appear highly complex. However, all machines, no matter how complex, are made up of one or more of the simple machines. The simple machines are:

lever, wheel and axle, pulley, inclined plane, wedge, screw, spring

Individually, each of these machines is a simple machine. When two or more simple machines are combined in such a way that they work as a single mechanism, the device is classified as a complex machine.

EXERCIS 6-2

A. Words and Phrases

1. effort
2. accomplishment
3. in detail
4. be converted from
5. convert ~ into
6. and so on
7. according to
8. at the same time
9. conversely
10. be classified as
11. be familiar with
12. torque
13. proportional
14. be made up of

B. Reading Comprehension

1. Put T or F to indicate if the sentences below are true or false according to the facts in the passage.

- Force is a measure of energy. ________________
- A concept that no machine can produce more work than the amount of work that is put into the machine is called as the conservation of energy principle. ________________
- Energy is found in many different forms, but it cannot be converted from one form to another. ________________
- When a machine produces an increase in force, the distance moved decreases proportionally. ________________
- The work done in a mechanical system is the product of force multiplied by the volume. ________________
- A system is said to possess force when it can work. ________________
- Kinetic energy is the energy that a body has as a result of its velocity. ________________

2. Answer the following questions.

- Define the work done in a mechanical system.
- Please explain the conservation of energy principle in more than two ways.
- Explain the difference of the simple machine and the complex machine.

bibliography

Katsuhiko Ogata, *System Dynamics*, pp. 84~95, Prentice-Hall, 1998.

R.A. Hinrichs & M. Kleinbach, *Energy its Use and the Environment*, pp. 1~3, Harcourt, 2002.

Katsuhiko Ogata, *System Dynamics*, pp. 84~95, Prentice-Hall, 1998.

Unit

07

Thermal Engineering

Unit 07 Thermal Engineering

1 Examples of Steady-State Processes

In this section, we consider a number of examples of steady-state processes in which there is one fluid stream entering and one leaving the control volume, such that the first law can be written in the form of Eq. 6.13. Some may instead utilize control volumes that include more than one fluid stream, such that it is necessary to write the first law in the more general form of Eq. 6.10.

Heat Exchanger

A steady-state heat exchanger is a simple fluid flow through a pipe or system of pipes, where heat is transferred to or from the fluid. The fluid may be heated or cooled, and may or may not boil, liquid to vapor, or condense, vapor to liquid. One such example is the condenser in an R-134a refrigeration system as shown in Fig 7.1. Superheated vapor enters the condenser, and liquid exits. The process tends to occur at constant pressure, since a fluid flowing in a pipe usually undergoes only a small pressure drop, because of fluid friction at the walls. The pressure drop may or may not be taken into account in a particular analysis. There is no means for doing any work (shaft work, electrical work, etc.), and changes in kinetic

Terminology

superheated vapor
fluid friction
shaft work
constant pressure
refrigeration system
pressure drop
control volume
boil
condenser
heat exchanger
steady−state process

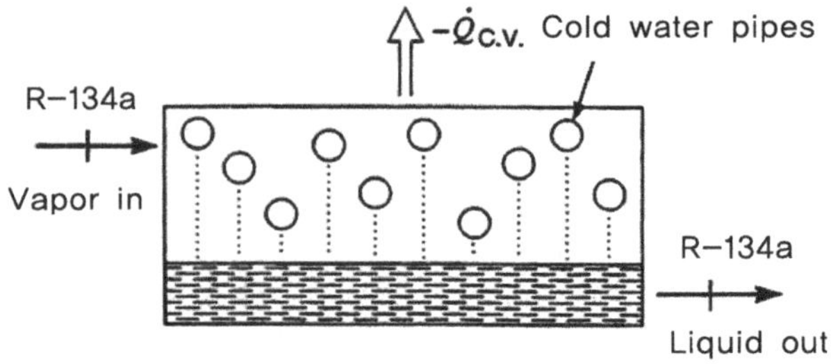

▲ **Figure 7.1** A refrigeration system condenser.

and potential energies are commonly negligibly small. (One exception may be a boiler tube in which liquid enters and vapor exits at a much larger specific volume. In such a case, it may be necessary to check the exit velocity using Eq. 7.3.) The heat transfer in most heat exchangers is then found from Eq. 6.13 as the change in enthalpy of the fluid. In the condenser shown in Fig 7.1, the heat transfer out of the condenser then goes to whatever is receiving it, perhaps a stream of air or of cooling water. It is often simpler to write the first law around the entire heat exchanger, including both flow streams, in which case there is little or no heat transfer with the surroundings. Such a situation is the subject of the following example.

Terminology

enthalpy
heat transfer
steady−state device
liquid−propellant
surrounding
cooling water
nozzle
combustion chamber

Nozzle

A nozzle is a steady-state device whose purpose is to create a high-velocity fluid stream at the expense of the fluid's pressure. It is contoured in an appropriate manner to expand a flowing fluid smoothly to a lower pressure, thereby increasing its velocity. There is no means to do any work−there are no moving parts. There is little or no change in potential energy and usually little or no heat transfer. An exception is the large nozzle on a liquid-propellant rocket, such as was described in Section 1.7, in which the cold propellant is commonly circulated around the outside of the nozzle walls before going to the combustion chamber, in order to keep the nozzle from melting. This case, a nozzle with significant heat transfer, is the exception and would be noted in such an application. In addition, the kinetic energy of the fluid at the nozzle inlet is usually small and would be neglected if its value is not known.

Diffuser

A steady-state diffuser is a device constructed to decelerate a high-velocity fluid in a manner that results in an increase in pressure of the fluid. In essence, it is the exact opposite of a nozzle, and it may be thought of as a fluid flowing in the opposite direction through a nozzle, with the opposite effects. The assumptions are similar to those for a nozzle, with a large kinetic energy at the diffuser inlet and a small, but usually not negligible, kinetic energy at the exit being the only terms besides the enthalpies remaining in the first law, Eq. 6.13.

Throttle

A throttling process occurs when a fluid flowing in a line suddenly encounters a restriction in the flow passage. This may be a plate with a small hole in it, as shown in Fig 7.2, it may be a partially closed valve protruding into the flow passage, or it may be a change to a much smaller

Terminology

throttling process
diffuser
flow passage

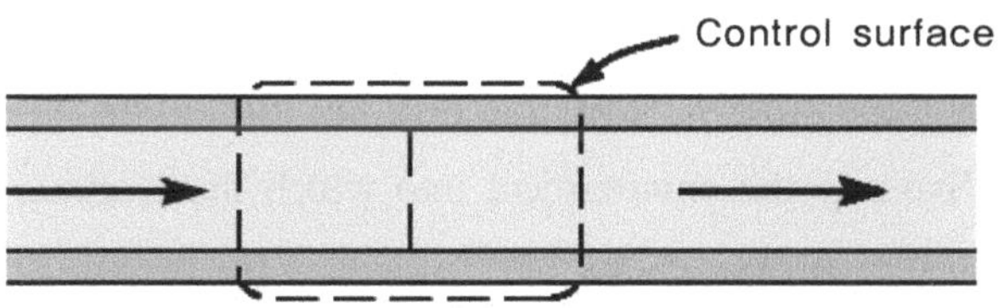

▲ **Figure 7.2** The throttling process.

diameter tube, called a capillary tube which is normally found on a refrigerator. The result of this restriction is an abrupt pressure drop in the fluid, as it is forced to find its way through a suddenly smaller passageway. This process is drastically unlike the smoothly contoured nozzle expansion and area change, which results in a significant velocity increase. There is typically some increase in velocity in a throttle, but both inlet and exit kinetic energies are usually small enough to be neglected. There is no means for doing work and little or no change in potential energy. Usually, there is neither time nor opportunity for an appreciable heat transfer, such that the only terms left in the first law, Eq. 6.13, are the inlet and exit enthalpies. We conclude that a steady-state throttling process is approximately a pressure drop at constant enthalpy, and we will assume this to be the case unless otherwise noted.

Frequently, a throttling process involves a change in the phase of the fluid. A typical example is the flow through the expansion valve of a vapor-compression refrigeration system. The following example deals with this problem.

Terminology

vapor−compression
expansion valve
exit
working fluid
steam turbine
atmosphere
phase
turbine
inlet
capillary tube

Turbine

A turbine is a rotary steady-state machine whose purpose is to produce shaft work (power, on a rate basis) at the expense of the pressure of the working fluid. Two general classes of turbines are steam (or other working fluid) turbines, in which the steam exiting the turbine passes to a condenser, where it is condensed to liquid, and gas turbines, in which the gas usually exhausts to the atmosphere from the turbine. In either type, the turbine exit pressure is fixed by the environment into which the working fluid exhausts, and the turbine inlet pressure has been reached by previously pumping or compressing the working fluid in another process. Inside the turbine, there are two distinct processes. In the first, the working fluid passes through a set of nozzles, or the equivalent−fixed blade passages contoured to expand the fluid to a lower pressure and to a high velocity. In the second process inside the turbine, this high-velocity fluid stream is directed onto a set of moving (rotating) blades, in which the velocity is reduced before being discharged from the passage. This directed velocity decrease produces a torque on the rotating shaft, resulting in a shaft work output. The low-velocity, low-pressure fluid then exhausts from the turbine.

The first law for this process is either Eq. 6.10. or 6.13. Usually, changes in potential energy are negligible, as is the inlet kinetic energy. Often, the exit kinetic energy is neglected, and any heat rejection from the turbine is undesirable and is commonly small. We therefore normally assume that a turbine process is adiabatic, and the work output in this case reduces to the decrease in enthalpy from the inlet to exit states. In the following example, however, we include all the terms in the first law and study their relative importance.

Compressor and pump

The purpose of a steady-state compressor (gas) or pump (liquid) is the same: to increase the pressure of a fluid by putting in shaft work (power, on a rate basis). There are two fundamentally different classes of compressors. The most common is a rotary-type compressor (either axial flow or radial/centrifugal flow), in which the internal processes are essentially the opposite of the two processes occurring inside a turbine. The working fluid enters the compressor at low pressure, moving into a set of rotation blades, from which it exits at high velocity, a result of the shaft work input to the fluid. The fluid then passes through a diffuser section, in which it is decelerated in a manner that results in a pressure increase. The fluid then exits the compressor at high pressure.

The first law for the compressor is either Eq. 6.10. or 6.13. Usually, changes in potential energy are negligible, as is the inlet kinetic energy.

Terminology

compressor
rotating shaft
adiabatic
pump
shaft work
potential energy
kinetic energy

Often the exit kinetic energy is neglected, as well. Heat rejection from the working fluid during compression would be desirable, but it is usually small in a rotary compressor, which is a high-volume flow-rate machine, and there is not sufficient time to transfer much heat from the working fluid. We therefore normally assume that a rotary compressor process is adiabatic, and the work input in this case reduces to the change in enthalpy from the inlet to exit states.

In a piston/cylinder-type compressor, the cylinder usually contains fins to promote heat rejection during compression (or the cylinder may be water-jacketed in a large compressor for even greater cooling rates). In this type of compressor, the heat transfer from the working fluid is significant and is not neglected in the first law. As a general rule, in any example or problem in this text, we will assume that a compressor is adiabatic unless otherwise noted.

Terminology

flow−rate
fin
rotary compressor
flow rate

EXERCIS 7-1

A. Words and Phrases

1. condense
2. vapor
3. superheat
4. take ~ into account
5. at the expense of
6. propellant
7. passage
8. throttle
9. drastically
10. appreciable
11. flow rate
12. contour
13. shaft
14. turbine
15. exhaust
16. expand
17. radial
18. centrifugal
19. result in
20. abrupt
21. deal with
22. equivalent

B. Reading Comprehension

1. Put T or F to indicate if the sentences below are true or false according to the facts in the passage.

- In a steady-state heat exchanger, pressure is considered to be constant. ______________
- A nozzle is a steady-state device obtaining high pressure at the expense of fluid's high velocity. ______________
- The function of a diffuser is very similar to that of a nozzle. ______________
- Condenser is one of the steady-state heat exchangers. ______________
- When a fluid flowing in a line suddenly encounters a restriction in the flow passage, a process called "throttling" occurs. ______________
- One of the characteristics of the throttling process is constant pressure at constant enthalpy. ______________

- The purpose of a turbine is to produce shaft work. ________________
- A throttling process does not involve any change in the phase of the fluid. ________________
- Changes in potential energy can be neglected in the steady-state turbine processes. ________________
- In general, the kinetic energy of the fluid at the nozzle exit is small compared with that at the nozzle inlet. ________________
- The purpose of a steady-state compressor is different from that of a steady-state pump. ________________
- Working fluids of a compressor are compressible gases. ________________

2. Answer the following questions.

- Write down the purpose of a nozzle.
- Explain why a nozzle does not do any work.
- What is the purpose of a steady-state diffuser?
- Is a nozzle the exact opposite of a diffuser? If yes, please explain.
- Write down the typical characteristics of the throttling process.
- When does a throttling process occur?
- By what is the turbine exit pressure fixed?
- Write down the terms which can be neglected in the energy equation of the first law during normal steady-state turbine processes.
- Can we assume a normal steady-state turbine process to be adiabatic? In this case what equals the amount of the work output?
- Why are the purposes of a steady-state compressor and a steady-state pump the same?
- List all the terms in the first law which are negligibly small or can be neglected during a steady-state process in a rotary-type compressor.

2) The Fundamental Concepts and the Basic Modes of Heat Transfer

It is customary to categorize the various heat transfer processes into three basic types or modes, although, as will become apparent as one studies the subject, it is certainly a rare instance when one encounters a problem of practical importance which does not involve at least two, and sometimes all three, of these modes occurring simultaneously. The three modes are conduction, convection, and radiation.

Heat conduction is the term applied to the mechanism of internal energy exchange from one body to another, or from one part of a body to another part, by the exchange of the kinetic energy of motion of the molecules by direct communication or by the drift of energy or heat passes from the higher energy molecules to the lower energy ones(i.e., from a high temperature region to a low temperature region). The distinguishing feature of conduction is that it takes place within the boundaries of a body, or across the boundary of a body into another body placed in contact with the first, without an appreciable displacement of the matter comprising the body.

A metal bar heated on one end will, in time, become hot at its other end. This is the simplest illustration of conduction. The laws governing conduction can be expressed in concise mathematical terms, and the analysis of the heat flow can be treated analytically in many instances.

Convection is the term applied to the heat transfer mechanism which occurs in a fluid by the mixing of one portion of the fluid with another portion due to gross movements of the mass of fluid. The actual process of energy transfer from one fluid particle or molecule to another is still one of conduction, but the energy may be transported from one point in space to another by the displacement of fluid itself. The fluid motion may be caused by external mechanical means (e.g, by a fan, pump, etc.), in which case the process

Terminology

heat conduction
mechanism
convection
radiation

is called forced convection. If the fluid motion is caused by density differences which are created by the temperature differences existing in the fluid mass, the process is termed free convection or natural convection. The circulation of the water in a pan heated on a stove is an example of free convection. The important heat transfer problems of condensing and boiling are also examples of convection—involving the additional complication of a latent heat exchange.

It is virtually impossible to observe pure heat conduction in a fluid because as soon as a temperature difference is imposed on a fluid, natural convection currents will occur as a result of density differences.

The basic laws of heat conduction must be coupled with those of fluid motion in order to describe, mathematically, the process of heat convection. The mathematical analysis of the resulting system of differential equations is perhaps one of the most complex fields of applied mathematics. Thus, for engineering applications, convection analysis will be seen to be a subtle combination of powerful mathematical techniques and the intelligent use of empiricism and experience. Thermal radiation is the term used to describe the electromagnetic radiation which has been observed to be emitted at the surface of a body which has been thermally excited. This electromagnetic radiation is emitted in all directions; and when it strikes another body, part may be reflected, part may be transmitted, and part may be absorbed. If the incident radiation is thermal radiation (i.e, if it is of the proper wavelength), the absorbed radiation will appear as heat within the absorbing body.

Thus, in a manner completely different from the two modes discussed above, heat may pass from one body to another without the need of a medium of transport between them. In some instances there may be a separating medium, such as air, which is unaffected by this passage of energy. The heat of the sun is the most obvious example of thermal radiation.

Terminology

free convection
natural convection
latent heat
density
forced convection
thermal radiation
electromagnetic radiation

There will be a continuous interchange of energy between two radiation bodies, with a net exchange of energy from the hotter to the colder. Even in the case of thermal equilibrium, an energy exchange occurs, although the net exchange will be zero.

Terminology

thermal equilibrium

EXERCIS 7-2

A. Words and Phrases

1. customary	2. in contact with	3. conduction
4. convection	5. external	6. latent
7. empiricism	8. electromagnetic	9. radiation
10. comprise	11. complication	12. emit
13. take place	14. illustration	15. concise
16. analytically	17. impose	18. be coupled with

B. Reading Comprehension

1. Put T or F to indicate if the sentences below are true or false according to the facts in the passage.

- Heat conduction is a mechanism of internal energy exchange from one body to another, or from one part of a body to another part. ______________
- Convection is a heat transfer mechanism which occurs both in a solid and in a fluid. ______________
- It is very easy to observe pure conduction phenomena in any fluid. ______________
- Free convection is caused by density differences created by the temperature differences existing in the fluid. ______________
- Condensing and boiling involve latent heat exchanges. ______________

2. Answer the following questions.

- What are the three modes of heat transfer? List them all.
- What is the distinguishing feature of conduction?
- What is the distinguishing feature of convection?
- What is the distinguishing feature of radiation?
- What are the two types of convection? List them.

bibliography

R.E. Sonntag & C. Borgnakke, *Introduction to Engineering Thermodynamics*, pp. 144~154, John Wiley & Sons, 2007.

A.J. Chapman, *Heat Transfer*, pp. 3~4, Macmillan, 1974.

Unit

08

Mechanisms

Unit 08 Mechanisms

1 The Study of Mechanisms

The study of mechanisms is very important. With the continuing advances made in the design of instruments, automatic controls, and automated equipment, the study of mechanisms takes on new significance. Mechanisms may be defined as that division of machine design which is concerned with the kinematic design of linkages, cams, gears, and gear trains. Kinematic design is design on the basis of motion requirements in contrast to design on the basis of strength requirements. An example of each of the mechanisms listed above will be given in order to present a comprehensive picture of the components to be studied.

A sketch of a linkage is shown in Fig 8.1. This particular arrangement is known as the slider-crank mechanism. Link 1 is the frame and is stationary, link 2 is the crank, link 3 is the connecting rod, and link 4 is the slider.

Terminology

slider-crank
linkage
mechanism
automatic control
cam
kinematic design
gear trains

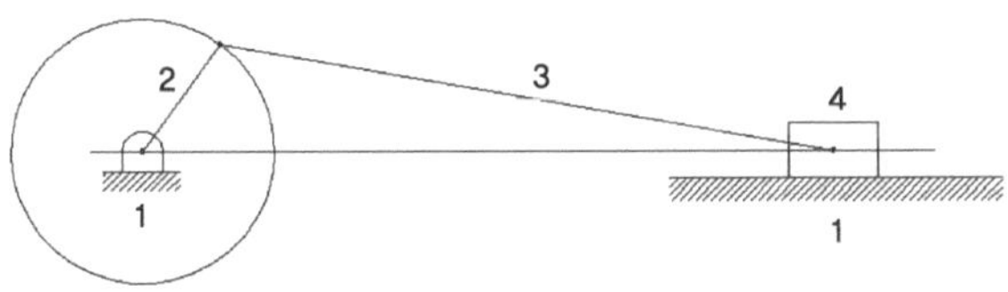

▲ **Figure 8.1** Slider-crank mechanism.

A common application of this linkage is in the internal-combustion engine where link 4 becomes the piston. This figure also demonstrates how difficult it may be to discern the basic kinematic device when looking at a photograph or a drawing of a complete machine. Figure 8.2 shows the kinematic diagram of the slider-crank mechanism corresponding to the left-side crankshaft – connecting-rod – piston. Such a kinematic diagram is much easier to work with and allows the designer to separate the kinematic considerations from the larger problem of machine design.

Figure 8.3 shows the sketch of a cam and follower. The cam rotates at a constant angular velocity, and the follower moves up and down. On the upward motion the follower is driven by the cam, and on the return motion it is driven by the action of gravity or of a spring. Cams are used in many machines, but one of the most common is the automotive engine where two cams are used per cylinder to operate the intake and exhaust valves. A three-dimensional cam is shown in Fig 8.4. In this cam, the motion of the follower depends not only upon the rotation of the cam but also upon the axial motion of the cam.

Gears are used in many applications to transmit motion from one shaft to another with a constant angular velocity ratio. Figure 8.5 shows several commonly used gears.

Terminology

gravity
exhaust
axial motion
angular velocity
automotive engine
follower
intake
internal–combustion
cylinder
crankshaft
engine

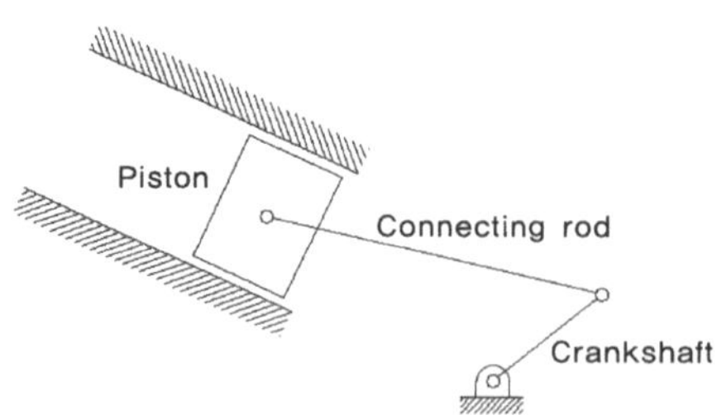

▲ **Figure 8.2** Kinematic diagram of engine mechanism.

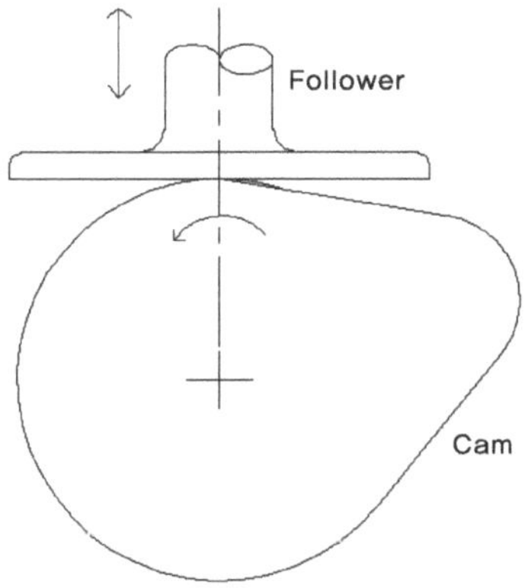

▲ **Figure 8.3**
Two-dimensional cam.

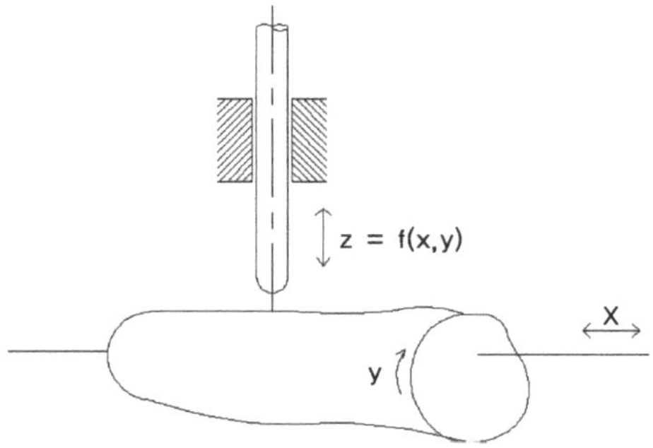

▲ **Figure 8.4**
Three-dimensional cam.

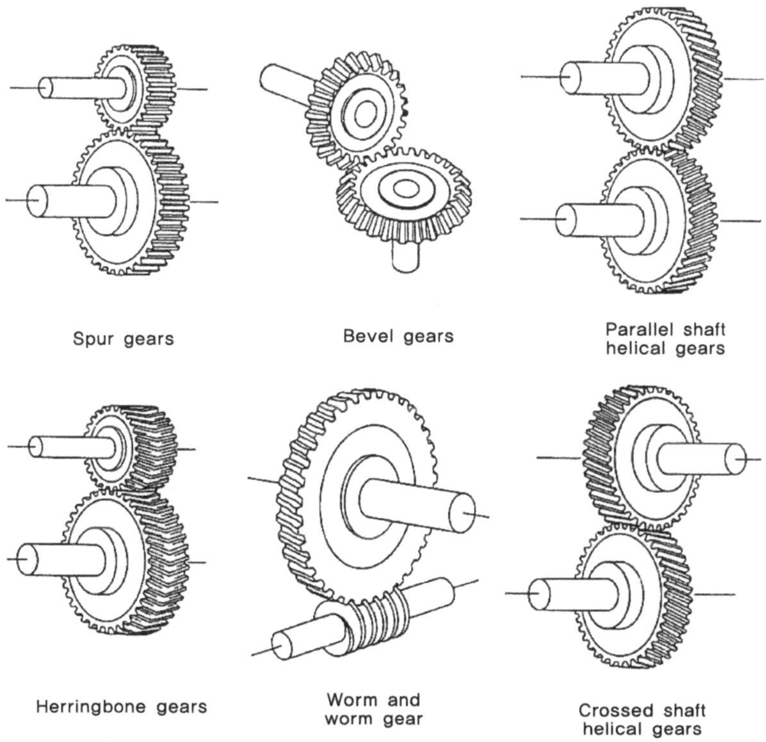

▲ **Figure 8.5**

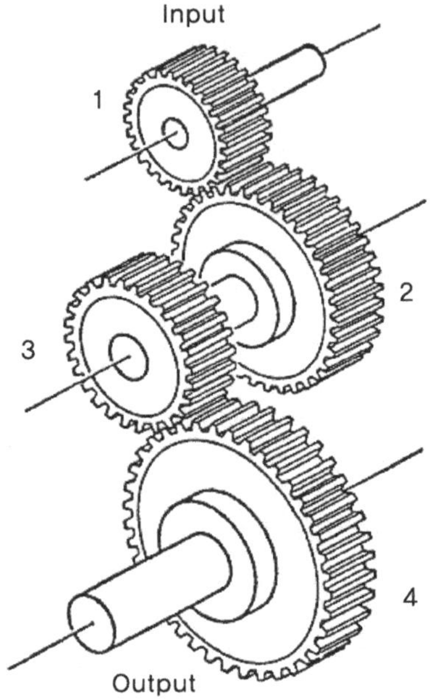

▲ **Figure 8.6** Gear train.

Terminology

transmission
gear train
automatic control

In some cases, the desired reduction in angular velocity is too great to achieve using only two gears. When this occurs, several gears must be connected together to give what is known as a gear train. Figure 8.6 shows a gear train where the speed is stepped down in going from gear 1 to gear 2 and again in going from gear 3 to gear 4. Gear 1 is the driver, and gears 2 and 3 are mounted on the same shaft. In many gear trains, it is necessary to be able to shift gears in and out of mesh so as to obtain different combinations of speeds. A good example of this is the automobile transmission where three speeds forward and one in reverse are obtained by shifting two gears.

In devices such as instruments and automatic controls, obtaining the correct motion is all-important. The power transmitted by the elements may be so slight as to be negligible, which allows the components to be proportioned primarily on the basis of motion, strength being of secondary importance.

There are other machines, however, where the kinematic analysis is only one step in the design. After it has been determined how the various machine components will act to accomplish the desired motion, the forces acting upon these parts must be analyzed. From this, the physical size of the parts may be determined. A machine tool is a good example; its strength and rigidity are more difficult to attain than the desired motions.

Terminology

kinematic analysis
machine component
machine tool

EXERCIS 8-1

A. Words and Phrases

1. be concerned with
2. in contrast to
3. comprehensive
4. discern
5. work with
6. proportioned
7. on the basis of
8. transmit
9. reduction
10. in and out of mesh
11. step down
12. forward
13. reverse
14. negligible

B. Reading Comprehension

1. Put T or F to indicate if the sentences below are true or false according to the facts in the passage.

- Kinematic design is design on the basis of strength requirement. ________
- In a three-dimensional cam, the motion of the follower depends wholly on the axial motion of the cam. ________
- Gears are used to transmit motion from one shaft to another. ________
- When the gear ratio is too large, three or more gears are used to achieve a desirable gear ratio. ________
- The physical size of the parts may be determined just after the kinematic analysis. ________
- It is very difficult to discern the basic kinematic device by looking at a photograph. ________
- A kinematic diagram allows the designer to separate the kinematic considerations from the larger problem of machine design. ________

2. Answer the following questions.

- What is the definition of mechanisms?
- Write down an example of the slider-crank mechanism?
- What is kinematic design?
- What is the name of the part which is driven by the cam?
- On what does the motion of the follower depend in a three-dimensional cam?
- What is the name of several gears connected together?
- What is the most important one in devices such as instrument and automatic controls?
- In designing machines, what is the next step after determining the kinematic design?

bibliography

Hamilton H. Mabie and Charles F. Reinholtz, *Mechanisms and Dynamics of Machinery*, pp. 1~5 John Wiley & Sons, 1987.

2) Kinematics and Dynamics of Machinery

In this text we will explore the topics of kinematics and dynamics of machinery in respect to the synthesis of mechanisms in order to accomplish desired motions or tasks, and also the analysis of mechanisms in order to determine their rigid-body dynamic behavior. These topics are fundamental to the broader subject of machine design. On the premise that we cannot analyze anything until it has been synthesized into existence, we will first explore the topic of synthesis of mechanisms. Then we will investigate techniques of analysis of mechanisms. All this will be directed toward developing your ability to design viable mechanism solutions to real, unstructured engineering problems by using a design process. We will begin with careful definitions of the terms used in these topics.

Kinematics and Kinetics

Terminology

Newton's second law
machinery
synthesis
design process
kinetic force
rigid−body

Kinematics	The study of motion without regard to forces.
Kinetics	The study of forces on systems in motion.

These two concepts are really not physically separable. We arbitrarily separate them for instructional reasons in engineering education. It is also valid in engineering design practice to first consider the desired kinematic motions and their consequences, and then subsequently investigate the kinetic forces associated with those motions. The student should realize that the division between kinematics and kinetics is quite arbitrary and is done largely for convenience. One cannot design most dynamic mechanical systems without taking both topics into thorough consideration. It is quite logical to consider them in the order listed since, from Newton's second law, F=ma, one typically needs to know the accelerations (a) in order to compute the dynamic forces

(F) due to the motion of the system's mass (m). There are also many situations in which the applied forces are known and the resultant accelerations are to be found.

One principal aim of kinematics is to create (design) the desired motions of the subject mechanical parts and then mathematically compute the positions, velocities, and accelerations which those motions will create on the parts. Since, for most earthbound mechanical systems, the mass remains essentially constant with time, defining the accelerations as a function of time then also defines the dynamic forces as a function of time. Stresses, in turn, will be a function of both applied and inertial (ma) forces. Since engineering design is charged with creating systems which will not fail during their expected service life, the goal is to keep stresses within acceptable limits for the materials chosen and environmental conditions encountered. This obviously requires that all system forces be defined and kept within desired limits. In machinery which moves (the only interesting kind), the largest forces encountered are often those due to the dynamics of the machine itself. These dynamic forces are proportional to acceleration, which brings us back to kinematics, the foundation of mechanical design. Very basic and early decisions in the design process involving kinematic principles can be crucial to the success of any mechanical design. A design which has poor kinematics will prove troublesome and perform badly.

Terminology

dynamic force
analog
resultant acceleration
stress
inertial force

Mechanisms and Machines

A mechanism is a device which transforms motion to some desirable pattern and typically develops very low forces and transmits little power. A machine typically contains mechanisms which are designed to provide significant forces and transmit significant power. Some examples of common mechanisms are a pencil sharpener, a camera shutter, and analog clock, a folding chair, an adjustable desk lamp, and an umbrella. Some examples of machines which possess motions similar to the mechanisms listed above are a food blender, a bank vault door, an automobile transmission, a bulldozer, a robot, and an amusement park ride. There is no clear-cut dividing line between mechanisms and machines. They differ in degree rather than in kind. If the forces or energy levels within the device are significant, it is considered a machine; if not, it is considered a mechanism. A useful working definition of a mechanism is a system of elements arranged to transmit motion in a predetermined fashion. This can be converted to a definition of a machine by adding the words force and energy after motion.

Mechanisms, if lightly loaded and run at slow speeds, can sometimes be treated strictly as kinematic devices; that is, they can be analyzed kinematically without regard to forces. Machines (and mechanisms running at higher speeds), on the other hand, must first be treated as mechanisms, a kinematic analysis of their velocities and accelerations must be done, and then they must be subsequently analyzed as dynamic systems in which their static and dynamic forces due to those accelerations are analyzed using the principles of kinetics.

Terminology

food blender
robot
vault door
kinematic device

EXERCIS 8-2

A. Words and Phrases

1. explore
2. premise
3. be directed toward
4. viable
5. unstructured
6. in respect to
7. without regard to
8. arbitrarily
9. subsequently
10. earthbound
11. be proportional to
12. be crucial to
13. troublesome
14. adjustable
15. amusement
16. clear-cut
17. predetermined
18. strictly
19. be charged with
20. convert

B. Reading Comprehension

1. Put T or F to indicate if the sentences below are true or false according to the facts in the passage.

- We must consider desired kinematic motions before kinetic forces. ________________
- All system forces must be defined and kept within desired limits. ________________
- There is a clear-cut dividing line between mechanisms and machines. ________________
- When the forces within the device are negligible, it is considered a machine. ________________
- Lightly loaded mechanisms with slow speeds can be treated as kinetic devices. ________________
- The largest forces encountered in some moving machinery are due to the dynamics of the machine itself. ________________

2. Answer the following questions.

- What is one of the principal aim of kinematics?
- What is the definition of a mechanism?
- What is the basic difference between mechanisms and machines?
- What is the basic difference between kinematics and dynamics?
- How does the author separate kinematics and kinetics in this paragraph?

bibliography

Robert L. Norton, *Design of Machinery*, pp. 3~4, McGraw-Hill, 1999.

Unit

09

Manufacturing Processes and Mechanical Vibrations

Unit 09 Manufacturing Processes and Mechanical Vibrations

1 Manufacturing Processes

Casting is the manufacturing process whereby the desired material is heated to the liquid state, then poured into a previously prepared mold cavity of proper design, and allowed to solidify in the mold before being extracted, trimmed and cleaned.

From this definition it is apparent that a study of the casting process must include the following areas: patterns and molding; melting and pouring (flow); solidification and cooling; cleaning and inspection. In considering the castability of specific material, the properties and relationships to be taken into account are:(1) the mold materials;(2) the solubility or chemical reactivity of the part material and the mold material (3) the solubility in the material of its atmosphere at low and high temperatures; (4) the thermal expansion and thermal conduction of both the part and mold materials; and (5) the cost and availability of both the part and mold materials.

There are many casting processes, which are generally classified by the type of mold used to make the casting. The more common of the processes as listed in order of decreasing application are: green sand, die, permanent mold, shell, investment, baked sand, plaster. These processes will be discussed in more detail later.

Terminology

solubility
shell
die
baked sand
molding
plaster
solidification
investment
mold cavity
casting
castability
permanent mold
patterns
melting

The manufacturing plant in which castings are made is called a *foundry*. In the past, the process of casting was called *founding*, and it is occasionally called that today. Individual foundries are specialized according to the molding processes they use and the materials they cast. Thus, some foundries make only gray iron castings in sand molds, others make only permanent cast iron, whereas others restrict themselves to die casting zinc or aluminum base alloys. Some die casting plants make only castings of plastics, which are frequently called injection moldings.

There are many definitions of welding in the technical literature, with some of them explanations rather than definitions. The definitions vary from the all-exclusive "joining together of two or more members" to the all-exclusive "joining together of two or more piece of metal by melting or fusing their contacting surfaces". The first definition is completely unsatisfactory to a student of the manufacturing processes because it does not exclude the joining operations of riveting, bolting, press fits, tying together with wire, or, if the author is permitted an extreme liberty, the bonding of a bride and groom by a marriage ceremony. The second definition is equally unsatisfactory since it does not include cold-pressure welding, recrystallization welding, welding with adhesives, and includes only part of the heterogeneous welding processes.

Terminology

die casting
foundry
injection molding
welding
recrystallization

A suitable definition of welding for an engineer might be: *Welding is the manufacturing process of joining two or more pieces of material together by utilizing only the fundamental attracting forces that hold the atoms of a solid into their fixed positions.* Since the memorizing of a definition is but the first short step toward understanding a subject (the application or use being the major step), the following discussion should clarify what welding is and is not.

Since all physics, chemistry, and materials textbooks describe sufficiently well the four fundamental bonding forces, it is assumed that the reader is already familiar with them and can now begin to apply this knowledge.

These forces are therefore simply listed here in order of decreasing bonding strengths: (1) covalent, (2) ionic, (3) metallic, and (4) molecular. Thus the joining of two pieces of plastic or the joining of two pieces of metal or ceramic by means of an "adhesive" is truly welding if the forces holding the pieces together are van der Waal's molecular attractive forces. Plastic materials which are used as adhesives are usually noncrystalline, so it is not appropriate to refer to grain boundaries when discussing them. In describing their growth from small individual units or particles into larger masses, we refer to the crosslinking and interlocking of chainlike molecules (Figure 4-1a) as well as the mutual attraction of dipole molecules.

Forming is the manufacturing process by which the size or shape of a part is changed by the application of forces that produce stresses in the part which are greater than the yield strength and less than the fracture strength of the material. The applied forces may be tensile, compressive, bending, shearing, or a combination of all.

According to our definitions of forming and machining presented in this text, the manufacturing operations called shearing, blank, and punching are not forming operations but rather they are machining operations. (The metal deformation during shearing and machining is very similar, as you can see by comparing Figure 5-1a and b.) The machine tools on which these operations are performed are generally the same as, or similar to, those used for the forming operations and not at all like the "conventional" metal cutting machine tools. Thus they are discussed in this chapter on forming, the place where they are traditionally discussed.

The number of different plastic working operations is very large and frequently they are associated with "trade" names. Many of the more common processes will be described in the following section under the title "equipment". All of the forming operations may be done either hot, cold, or warm; the distinction being made on the basis of the temperature at which the material is worked.

Terminology

bending
fracture strength
tensile
dipole
yield strength
blank
punching
machining
shearing
forming
plastic working
noncrystalline
mutual attraction

A forming operation is *hot-working* if the temperature of the material during working is above the recrystallization temperature of the material being worked, and it is *cold-working* if the temperature of the material is below its recrystallization temperature. A forming operation is referred to as warm-working if the temperature of the material is significantly above room temperature, but below the material's recrystallization temperature.

Machining is the manufacturing process in which the size, shape, or surface properties of a part are changed by *removing* the excess material. This is in contrast to forming where the materials is simply *moved*. Machining may be accomplished by locally straining the material to fracture through the relative motion of a tool and the work piece, as is done on the conventional processes. However, the material removal mechanism may be chemical, electrical or thermal as in the case of the unconventional processes such as chem-milling electrolytic, or electro-arc machining, respectively. In this light, flame cutting is a chemical machining process.

Machining is a relatively expensive process that should be specified only when high accuracy and good surface finish are required, which, unfortunately, is necessary for almost all mating parts. Machining today is almost exclusively performed by means of *machine tools*, which are power-driven machines that have provisions to hold and move the work piece as well as to hold and move a *cutting tool*. These new terms will all be described briefly.

Terminology

chem-milling
warm-working
hot-working
cold-working
electro-arc machining
recrystallization
machine tool
cutting
electrolytic

EXERCIS 9-1

A. Words and Phrases

1. solidify
2. solubility
3. availability
4. foundry
5. injection
6. heterogeneous
7. clarify
8. covalent
9. adhesive
10. crosslink
11. interlock
12. dipole
13. fracture
14. traditional
15. be associated with ~
16. recrystallization
17. refer
18. mate
19. flame

B. Reading Comprehension

1. Put T or F to indicate if the sentences below are true or false according to the facts in the passage.

- The manufacturing plant in which forming is made is called a foundry. ____________
- An appropriate definition of welding to a student is joining together of two or more members. ____________
- Joining two or more members using adhesives is not included in welding process. ____________
- Punching process is a machining operation. ____________
- If the temperature of the material during working is significantly above the room temperature, the plastic working operation is called warm-working. ____________
- Flame cutting cannot be classified as a machining process. ____________

- Machining is a relatively expensive process compared with other processes. ____________

2. Answer the following questions.

- Write a suitable definition of welding for an engineer.
- What are the four fundamental bonding forces used in welding processes?
- Write a suitable definition of casting for an engineer.
- Write a suitable definition of forming for an engineer.
- List the seven casting processes classified by the type of mold.
- List three kinds of plastic working operations according to their working temperatures?

bibliography

J. Datsko, *Materrial Properties and Manufacturing Processes,* pp. 102~103, John Wiley & Sons, 1966

2 Machining Processes

Of all the manufacturing processes that can be applied to the shaping and forming of raw materials into useful products, machining processes will always remain among the most important. The fundamental cutting processes in machining, those of bringing the work into contact with a cutting tool, are still very much in evidence and will always remain mainstays of the industry. What has changed is the way in which these processes are applied, the cutting tool materials, material used for products, and the methods of material removal. These new methods are quite different from the classical chip-producing processes. They include the use of lasers, electrical energy, electrochemical processes, ultrasound, high-pressure water jets, and high-temperature plasma arcs as material removal tools.

Computers have enhanced automated manufacturing; machining is no exception. Numerical control(NC) of machine tools has been available for many years. Today, typical machine tools of all types are equipped with their own computer numerical controls(CNCs) and almost every machining process can now be efficiently automated with an exceptional degree of accuracy, reliability, and repeatability. In fact, machining processes have become so sophisticated and reliable that the human operator of a machine tool may now be replaced with a robot. This application of computer-driven automated equipment will profoundly affect the employment of machinists and machine operators in future years.

The computer has found its way into almost every other phase of manufacturing as well. One important area is computer-aided design(CAD). The age of drawing board design is fast drawing to a close. Design is now done on computer terminals and manufacturing equipment control programs are generated directly in computer-aided manufacturing(CAM), or fully computer-aided design and manufacturing(CAD/CAM), or computer-integrated manufacturing(CIM).

Terminology

forming
electrochemical
manufacturing process
plasma arcs
machining process
machine tools
CAM
CAD
NC
ultrasound
shaping
laser

The ultimate outcome of manufacturing automation will be an entirely automated factory in which material will be automatically transported between manufacturing cells that will perform similar manufacturing or machining operations. Such a system, called a flexible manufacturing system(FMS), will be fully computer controlled and already has become an industry standard.

Terminology

FMS
CIM
automated factory

EXERCIS 9-2

A. Words and Phrases

1. mainstay	2. removal	3. enhanced
4. be equipped with	5. accuracy	6. reliability
7. repeatability	8. sophisticated	9. reliable
10. be replaced with	11. profoundly	13. ultimate
14. in evidence	15. profound	16. draw to a close

B. Reading Comprehension

1. Put T or F to indicate if the sentences below are true or false according to the facts in the passage.

- In future years, machinists and machine operators will be reduced because of the application of computer-driven automated equipments. ________________
- An entirely automated factory which is fully computer controlled is called a flexible manufacturing system. ________________
- The processes which use ultrasound or high-pressure water jets are not in the boundary of machining. ________________
- Although machining processes have become sophisticated, a robot cannot replace the human operator. ________________

2. Answer the following questions.

- What is the main point the author want to say in the above passage?
- Explain the flexible manufacturing system(FMS).
- List new methods of machining in contrast to the classical ones.

bibliography

Richard R. Kibbe, John E. Neely, Roland O. Meyer, and Warren T. White, *Machine Tool Practices*, pp. 1~2, Prentice-Hall, 1995.

3 Mechanical Vibrations

All matter — solid, liquid and gas — is capable of vibration, e.g. vibration of gases occurs in tail ducts of jet engines causing troublesome noise and sometimes fatigue cracks in the metal. Vibration in liquids is almost always longitudinal and can cause large forces because of the low compressibility of liquids, e.g., pipes conveying water can be subjected to high inertia forces(or "water hammer") when a valve or tap is suddenly closed. Excitation forces caused, say, by changes in flow of fluids or out-of-balance rotating or reciprocating parts, can often be reduced by attention to design and manufacturing details. A typical machine has many moving parts, each of which is a potential source of vibration or shock-excitation. Designers face the problem of compromising between an acceptable amount of vibration and noise, and costs involved in reducing excitation.

Practically every object or system vibrates freely on its own after some initial disturbance. Mechanical excitation is a necessity for the production of sound—"musical" or "noise"—the audible range being about 18 Hz to 18 kHz. All musical instruments use vibrations caused by different excitations, e.g. drums and cymbals are struck, and harp and guitar strings are plucked—i.e. impulse excitation; Violin and cello strings are bowed—i.e. 'stick-slip' excitation; wind instruments and organ pipes are blown—i.e. wind excitation. In contrast, whilst it is not easy to detect and measure the motion of, say, an engine part vibrating with an amplitude of a fraction of a millimeter, it is relatively easy in the case of musical instruments.

Engineers use vibrations deliberately in such operations as the drilling and cutting of square holes, etc, but in most cases the source and cause of vibration is sought in order to reduce or eliminate it if possible, otherwise high local stresses may be induced causing eventual failure of material by fatigue. Hence, the importance of periodic inspection of parts (e.g crack-

Terminology

shock−excitation
excitation
vibration
inertia force
disturbance
water hammer
stick−slip

detection) where fatigue is a possibility must be acknowledged

Fortunately, the sine (or cosine) wave can be the basis of analysis of vibrations and oscillatory quantities because, however complicated or peculiar the shape of a periodic wave-form, it can always be built up as an assembly of suitable sine waves.

The mechanical vibrations dealt with are either excited by steady harmonic forces (i.e obeying sine and cosine laws in cases of forced vibrations) or, after an initial disturbance, by no external force apart from gravitational force called weight (i.e in cases of natural, autonomous or free vibrations). Harmonic vibrations are said to be 'simple' if there is only one frequency as represented diagrammatically by a sine or cosine wave of displacement against time.

Vibration of a body or material is periodic change in position or displacement from a static-equilibrium position. Associated with vibration are the inter-related physical quantities of acceleration, velocity and displacement -e.g. an unbalanced force causes acceleration (a = F/m) in a system which, by resisting, induces vibration as a response.

We shall see that vibratory or oscillatory motion may be classified broadly as (i) transient, (ii) continuing or steady-state and (iii) random.

Terminology

transient
gravitational force
harmonic force
oscillatory
fatigue
unbalanced force
periodic wave
equilibrium position
random
steady-state

EXERCIS 9-3

A. Words and Phrases

1. troublesome
2. longitudinal
3. compressibility
4. reciprocating
5. compromising
6. deliberately
7. eliminate
8. autonomous
9. induce
10. whilst
11. pluck
12. apart from
13. vibratory
14. oscillatory

B. Reading Comprehension

1. Put T or F to indicate if the sentences below are true or false according to the facts in the passage.

- Every object or system vibrates freely on its own after initial disturbance. ____________
- Vibration in gases can cause large forces because of its high compressibility. ____________
- Violin and cello strings are struck to generate sound. ____________
- Periodic inspection is important where fatigue is a possibility. ____________
- When steady excitation forces applied to a system, only natural vibrations are generated. ____________
- Designers must eliminate sources of vibration completely. ____________
- It is relatively easy to detect sound in the case of musical instruments. ____________

2. Answer the following questions.

- Classify vibratory motion into three categories.
- What is the audible range of sounds to human ears according to the paragraph?

bibliography

A.C. Walshaw, *Mechanical Vibrations with Applications*, pp. 21~22, Ellis Horwood Limited, 1984.

Unit

10

Air Conditioning

Unit 10 Air Conditioning

1) The Kyoto Protocol on Climate Change

In December 1997, 167 nations met in Japan, under the auspices of the United Nations, to forge what is known as the "Kyoto Protocol." This document is the first international attempt to place legally binding limits on greenhouse gas emissions from developed countries. The Protocol set as a goal to cut by 2008-2012 the combined emissions of greenhouse gases from developed countries by about 5% from their 1990 levels. However, the Kyoto Protocol does not set any binding limits on emissions from developing countries. The United States, which emits the most CO_2 of any country, and which also has the highest per capita CO_2 emissions, will need to meet a reduction goal of 7% less than 1990. The U.S. Department of Energy estimates that U.S. carbon emissions by 2010 will increase by 34% in the absence of any change in energy policy or consumer behavior.

Since about 83% of greenhouse gas emissions in 1990 were CO_2 released by the use of fossil fuels, actions taken to reduce these emissions are likely to have a significant impact on energy markets.

Increases in energy prices may be required to meet these limits. It is expected that the cost of electricity generated by coal will rise and the share of electricity generated by natural gas and renewables will increase.

Terminology

carbon
fossil fuel
greenhouse gas
emission

EXERCIS 10−

A. Words and Phrases

1. auspice
2. protocol
3. emission
4. capita
5. policy
6. fossil
7. be likely to
8. share
9. renewable

B. Reading Comprehension

1. Put T or F to indicate if the sentences below are true or false according to the facts in the passage.

- According to the goal set by "Kyoto Protocol", developing countries must reduce the combined emissions of greenhouse gases by about 5% from their 1990 levels. ____________
- The Kyoto Protocol assigned some reductions in CO2 emissions to developing countries. ____________
- Fossil fuels are the main cause of greenhouse gas emission. ____________

2. Answer the following questions.

- How many nations were there in Japan to attend the meeting to forge the "Kyoto Protocol"?
- What is the CO2 reduction rate assigned to the U.S. by the Kyoto Protocol?
- According to the paragraph, what is expected in energy prices by the Kyoto Protocol?

2) Air Conditioning

Definition and History. Air conditioning is the simultaneous control of the temperature, humidity, motion, and purity of the atmosphere in a confined space. Credit for first using the expression is given to S. W. Cramer, who presented a paper on humidity control in textile mills before the National Cotton Manufacturers' Association in 1907. The term "yarn conditioning" was in common usage, and "air conditioning" seemed to be a logical expression for the process of controlling the atmospheric conditions.

When most people hear the expression "air conditioning," they think of a cool temperature in a theater or restaurant in the summer and generally regard it as a development of the 1930's. However, by definition, air conditioning applies in the heating season as well as in the cooling season, and the air-conditioning industry really obtained its start in industrial applications in the early part of this century. Since the refrigerating engineer is more interested in summer air conditioning than in winter air conditioning, the discussion in this text will be concerned mainly with the former.

Terminology

air conditioning

Summer air conditioning includes cooling, which may be accomplished by one of several means. The best method for a given application depends upon the local atmospheric conditions; the quantity, temperature, and cost of water available; the type and cost of steam, gas, or electric power available; and the nature of the application itself. The following are the most common types of air-cooling systems;

1. Direct-refrigeration system, in which the refrigerant evaporator is an air-cooling unit placed so that the air to be cooled passes directly over it.
2. Indirect refrigeration system, in which the refrigerant evaporator cools a circulating medium, such as water or brine, which in turn is used

in the air-cooling unit.

3. Well-water system circulating the water through the air-cooling unit.
4. Combination of refrigeration and well-water system.
5. Evaporative cooling system, in which the air-cooling unit is an air washer or other device presenting a large amount of wetted surface in the air stream and the water used is recirculated. Only enough fresh water is added to make up for that which evaporates and to flush out the system periodically.

In addition to the type of cooling equipment to use, the method of air distribution employed must be given careful consideration. Often the latter is the most critical part of the installation. Industrial applications may require an even temperature and air movement throughout. Disagreeable drafts must be avoided in comfort applications. Satisfactory methods of supplying air from either the ceiling or side wall are known. If commercial outlets are to be used, reliable data on them are available from several manufacturers. In considering the duct-sizing methods used, experience has indicated that less field adjustment and trouble are experienced when the static regain method is used.

Terminology

evaporative cooling

EXERCIS 10-

A. Words and Phrases

1. textile
2. mill
3. yarn
4. evaporative
5. make up for ~
6. flush
7. distribution
8. draft
9. regain
10. outlet

B. Reading Comprehension

1. Put T or F to indicate if the sentences below are true or false according to the facts in the passage.

- S.W. Cramer first used the expression 'air conditioning'. ________________
- The refrigerating engineer is more interested in winter air conditioning than in summer. ________________
- Even temperature and air movement is very important in industrial applications. ________________

2. Answer the following questions.

- Define the air conditioning according to the paragraph above.
- What do the most people think of when they hear the expression "air conditioning"?
- What do you have to consider to accomplish a comfortable summer air conditioning?
- Explain the difference between a direct refrigeration system and an indirect refrigeration system.
- Explain the evaporative cooling system.

3 Metabolism and comfort

The human body differs from the insentient machine in tolerating only a narrow band of temperature(36℃ to 38℃) and in having an appreciation of comfort. Defining the latter is not easy because of its subjective nature: in the long run a person's comfort can only be assessed by posing the question, 'Are you comfortable?' Even then the individual may have difficulty in phrasing a satisfactory answer. Establishing a numerical scale of comfort, interpreted from simple physical measurements, has not proved straightforward although Fanger(1972) has proposed a comfort equation (see section 4.7) that allows the determination of thermal comfort in terms of activity, clothing, dry-bulb temperature, the mean radiant temperature of the surrounding surfaces, relative humidity and air velocity.

Terminology

metabolism
muscular contraction

The body eats and digests food which, through the intake of oxygen by breathing, is converted into energy for useful work with the liberation of heat and the emission of waste products. The chemical changes occurring in the liver and other parts of the body, and muscular contraction, release thermal energy which is transported throughout the body by the circulation of the blood so that it can be lost to the outside world and the temperature of the deep tissues kept at comparatively constant value (about 37.2℃) if good health is to be preserved. The efficiency of the body—defined as W/(M-44), where W is the rate of working, M the metabolic rate of energy production for the particular activity and 44 W m^{-2} is the basal metabolic rate per unit area of bodily surface—is not very great, varying from 0 per cent when at rest to a maximum of about 20 per cent when walking up a one in four gradient at 10 $km \cdot h^{-1}$. So most of the energy produced by the bodily metabolism must be dissipated as heat to the environment.

In good health, the thermo-regulatory system of the body exercises automatic control over the deep tissue temperature by establishing a correct thermal balance between the body and its surroundings. Apart from the matter of good health the comfort of an individual also depends on this balance and it might therefore be interpreted as the ease with which such a thermal equilibrium is achieved. ASHRAE Comfort Standard 55-74 goes further and defines comfort as '... that state of mind which expresses satisfaction with the thermal environment...' but points out that most research work regards comfort as a subjective sensation that is expressed by an individual, when questioned, as neither slightly warm nor slightly cool.

Terminology

thermal equilibrium

EXERCIS 10–

A. Words and Phrases

1. metabolism
2. insentient
3. appreciation
4. assess
5. phrasing
6. liver
7. basal
8. bodily
9. dissipate
10. sensation

B. Reading Comprehension

1. Put T or F to indicate if the sentences below are true or false according to the facts in the passage.

- The human body can tolerate a broad band of temperature. ________
- The thermal energy generated by metabolism is transported through the body by the circulation of the blood. ________
- Most energy produced by bodily metabolism must be dissipated as mechanical one to the environment. ________
- Food is converted into energy when it is oxidized inside the human body. ________
- The efficiency of the human body can rise up to 44%. ________

2. Answer the following questions.

- In terms of what can you determine the thermal comfort using the equation proposed by Fanger?
- Write down the efficiency of the human body using the rate of working(W) and the metabolic rate of energy production for the particular activity(M).
- What is the temperature of the deep tissues in a healthy human body?
- How does the ASHRAE Comfort Standard 55–74 define the term 'comfort'?

4 Pneumatic Control Designer

The technicians at Johnson Controls Company are very familiar with the benefits of the new designs for energy-efficient office buildings. After all, they install and service the control systems that make them work. Energy-efficient buildings are *tighter* than buildings have ever been before; they don't leak. You can't open windows in energy-efficient buildings. Exterior doorways are revolving, or include a foyer, so that air will not rush in when doors are opened. Leaky buildings make for cold drafts in the winter and muggy conditions in the summer, and the heating and air-conditioning systems have to work overtime to make the corrections.

But buildings are occupied by people who need fresh air. Tight buildings don't supply fresh air; therefore, it must be introduced into the hot-air and cold-air ducts in a regulated amount. This is called *outside air* and typically represents ten percent of the air that is recirculated through the warm-air and cold-air duct system. Johnson Controls is an international company that specializes in the pneumatic controls that regulate the outside air supply. Pneumatics, as it applies to outside air control, opens and closes dampers in the outside air ducts in a variable manner to conserve the energy necessary to condition the air. The dampers move by the action of a pushrod attached to an air-operated piston and cylinder, called a damper motor, that are opposed by a spring so that the greater the pressure in the cylinder, the further the rod will open the damper. The technicians that make up the workforce of the company in Johnson Controls' locations in major cities are often engineering technicians who are well versed in the field of thermodynamics.

Terminology
pneumatic control
thermometer

In Johnson's systems, thermometers use air to measure temperature. In a pneumatic thermometer (Fig 4-10), which could be easily converted to a pneumatic thermostat, the thermometer does not record the temperature of a substance directly, as would a mercury-filled thermometer. Rather, the sensed temperature creates a unique pressure in an air-filled sensing bulb that is displayed on a pressure gauge. As the temperature of the air increases due to the placement of the probe in a hot environment, the pressure builds to move a needle in a pressure gauge. The pressure gauge is calibrated directly in terms of temperature.

Terminology

probe
pressure gauge

EXERCIS 10–

A. Words and Phrases

1. pneumatic 2. foyer 3. muggy
4. damper 5. thermostat 6. mercury
7. probe 8. gauge

B. Reading Comprehension

1. Put T or F to indicate if the sentences below are true or false according to the facts in the passage.

- You cannot open windows in energy-efficient buildings. ________
- A pneumatic thermometer records the temperature of a substance directly. ________
- Typically, the amount of fresh air admitted from the outside is ten percent of the recirculated air in energy-efficient buildings. ________
- You don't need any air duct in energy–efficient buildings. ________

2. Answer the following questions.

- How is the pneumatics used to outside air control?
- Explain the principle of a pneumatic thermometer.

bibliography

R.A. Hinrichs & M. Kleinbach, *Energy its Use and the Environment*, pp. 28, Harcourt, 2002.

R.C. Jordan & G,B Priester, *Refrigeration and Air Conditioning*, pp. 444~445, Prentice-Hall, 1974.

W.P. Jones, *Air Conditioning Engineering*, pp. 80, Butterworth-Heinemann, 2001.

J. Kamm, *Heat & Power Thermodynamics*, pp. 75, Delmar, 1997.

Unit 11

Electricity and Electronics

Unit 11 Electricity and Electronics

1 Basic Electricity

Defining Basic Electrical Terms

An electrical circuit can be defined as an electrical path between two or more devices for the purpose of carrying electrical current to produce some useful function. There are three important terms associated with an electrical circuit:

1. Voltage : the force that causes electrical current flow in the circuit, measured in units called volts. The volt is defined as the force required to produce 1 ampere of current through a resistance of 1 ohm. The voltage of battery produces current that flows in one direction and is referred to as direct-current (dc) voltage. It must be remembered that voltage is the force that produces current flow. Voltage does not flow but is the force that moves electrons in a conductor. Voltage is always in parallel or "across" a circuit component. Voltage is indicated by V or E.

2. Resistance : the opposition to current flow of the components in the circuit measured in units called ohms. The ohm, symbolized by the Greek capital letter omega (Ω), is defined as the value of electrical resistance that will allow 1 ampere of current to flow through a material

Terminology

resistance
electrical current
voltage
ampere
conductor
electrical circuit
dc
ohms
volts
resistance
electron

when 1 volt is applied. The conducting wires are assumed to have no, or zero, resistance, whereas the device in a circuit may have from several up to millions of ohms of resistance. Remember, resistance is stationary and does not move. It limits the amount of electrons flowing in a circuit. Resistance is indicated by R.

3. Current : the flow of electrons through the circuit caused by the force of the voltage, measured in units called amperes. The ampere is defined as the current flow produced when 1 volt is applied across 1 ohm of resistance. It is also defined that a current of 1 ampere is produced when 1 coulomb (6.25×10^{18} electrons) flows past a given point in 1 second. Remember, current does the flowing "in" or through the conductors and devices of a circuit. Current is indicated by I.

Resistance and Ohm's Law

Terminology
charge

The two fundamental quantities of voltage and current are related by a third quantity of equal importance: resistance. In any electrical system the pressure is applied voltage, and the result(or effect) is the flow of charge or current. The resulting level of current is controlled by the resistance of the system. The greater the resistance, the less the current, and vice versa. This effect is immediately obvious when we examine the most fundamental law of electric circuits: ohm's law.

$$I = \frac{E}{R}$$

where I = amperes (A)
E = volts (V)
R = ohms (Ω)

Note that the symbol for resistance is the capital letter R, measured in ohms, and that the Greek letter omega (Ω) is used as the symbol for its unit of measurement. The graphic symbol for the resistor appears in Fig 2.17 with the defined polarities for the indicated current direction (conventional flow). In words, it states that the high side (+) of a potential drop across a resistor is defined by that terminal through which the current enters the resistor.

Terminology

polarity
potential drop

EXERCIS 11–

A. Words and Phrases

1. be referred to as
2. opposition
3. stationary
4. the greater ~, the less ~
5. vice versa
6. immediately
7. obvious
8. conventional

B. Reading Comprehension

1. Put T or F to indicate if the sentences below are true or false according to the facts in the passage.

- Resistance is the force that causes electrical current flow in the circuit. ________________
- Current is the flow of electrons through the circuit caused by the force of the voltage. ________________
- Direct-current voltage produces current that flows in one direction. ________________
- Voltage is stationary and does not move. ________________
- Current is indicated by I. ________________
- In electrical systems, the pressure is applied current. ________________
- The symbol for resistance is R. ________________
- The conducting wires may have millions of ohms of resistance. ________________

2. Answer the following questions.

- What is the definition of voltage?
- What is the definition of current?
- What is the definition of resistance?
- Explain the Ohm's law.
- What are three important terms associated with an electrical circuit?

bibliography

Fredrick W. Hughes, *Fundamental DC/AC Circuits*, pp. 46, Prentice-Hall, 1990.

2) Electricity and Materials

Conductors

Materials that offer very little resistance (opposition) to current are called conductors. Copper, aluminum, and silver are good conductors. They have very low resistance. In general, those elements which have three or fewer electrons in the valence shell can be classified as conductors. However, even within those elements classified as conductors, there is wide variation in the ability to conductors, there is wide variation in the ability to conduct current. For example, iron has nearly six times the resistance of copper, though they are both considered conductors. Although silver is a slightly better conductor than copper, it is too expensive for common use. Aluminum is not as good a conductor as copper, but it is cheaper and lighter. Large aluminum conductors are used to bring electric energy into homes.

Superconductivity is the condition in which a material has no resistance. For many years, superconductivity could be demonstrated only at temperatures close to absolute zero, which is about -273℃ (degree Celsius) or -460°F (degrees Fahrenheit). Further research led the development of new material that exhibit superconductivity at temperatures well above absolute zero. The aim has always been to find materials that will have superconductivity at room temperature. Such materials would, obviously, greatly improve the efficiency of any electric system in which they could be used.

Terminology

conductor
copper
Celsius
free electron
insulator
silver
superconductivity
electric energy
aluminum
current carrier
Fahrenheit
valence shell

Insulators

Materials that offer a high resistance to current are called insulators. Even the best insulators do release an occasional free electron to serve as a current carrier. However, for most practical purposes we can consider an insulator to be a material that allows no current to flow through it. Common insulator materials used in electric devices are paper, wood, plastics, rubber, glass, and mica. Notice that common insulators are not pure elements. They are materials in which two or more elements are joined together to form a new substance. In the process of joining together, elements share their valence electrons. This sharing of valence electrons is called covalent bonding. It takes a lot of added energy to break an electron free of a covalent bond.

Semiconductors

Between the extremes of conductors and insulators are a group of elements known as semiconductors. Semiconductor elements have four valence electrons. Two of the best-known semiconductors are silicon and germanium. Semiconductors are neither good conductors nor good insulators. They allow some current to flow, yet they have a considerable amount of resistance. Semiconductors are extremely important industrial materials. They are the materials from which electronic devices such as transistors, integrated circuits (ICs), and solar cells are manufactured.

Terminology

germanium
silicon
solar cells
covalent bond
electric device
valence electron
semiconductor
mica
transistors
electronic devices
ICs

EXERCIS 11–

A. Words and Phrases

1. classified as
2. variation
3. close to
4. efficiency
5. release
6. occasional
7. be joined together
8. best–known
9. neither ～ nor ～
10. considerable
11. valence

B. Reading Comprehension

1. Put T or F to indicate if the sentences below are true or false according to the facts in the passage.

- Aluminum is a good conductor. ______________
- Semiconductor is a good insulator. ______________
- Common insulator is made of pure elements. ______________
- Aluminum is cheaper and lighter than copper. ______________
- Superconductivity is generated well above 0℃. ______________
- You need a lot of added energy to break an electron free of a covalent bond. ______________
- Transistors are made of semiconducting materials. ______________

2. Answer the following questions.

- What is semiconductors? Please explain.
- What is the practical purpose of an insulator?
- List six common insulator materials.

- What is the definition of conductors?
- What is the definition of superconductivity?

bibliography

Richard J. Fowler, *Electricity Principles and Applications*, pp. 25~26, McGraw-Hill, 1999.

3 Electronics

Electronics — to most of us this brings to mind a variety of things from "chips" and computers to television and transistors. Yet, while we agree on specific items that constitute electronics, its definition is elusive. In the next several paragraphs and the remainder of this prologue, we define electronics as used in this book not in the dictionary sense, but in a manner which attempts to convey the flavor and nuance of the discipline. We have chosen history as the vehicle to accomplish this because it is the efforts of individuals who contribute and have contributed to the field that truly define the discipline.

Electronics, in the strictest sense, is the science and technology of the motion of charges in a gas, vacuum, or semiconductor. Note that charge motion confined to a metal is not considered electronics. This was an historical division used early in the twentieth century to separate the already flourishing field of electrical engineering from the new and emerging field of electronic engineering. At that time electrical engineering dealt with devices that depended solely on the motion of elcctronic electrons in metals, such as motors, generator, light bulbs, and wire communication systems (telephone and telegraph). As we approach the end of the twentieth century, however, the historical division between electrical and electronic engineering no longer serves its original function.

Terminology

electronics
vacuum
wire communication
generator
light bulbs
electrical engineering
motors

Today practicing electrical engineers perform diverse function (design, development, production, research, and even teaching) with varied applications. They deal with systems by which we can communicate with one another worldwide, by which vast quantities of data are manipulated, and by which highly complex manufacturing processes are automated and with the elements used to realize them. The province of electrical engineering also includes the devices, circuits, and systems used for the generation, distribution, and conversion of electric energy. The group mentioned in the first of the two previous sentences possesses the common property of processing information; the group mentioned in the second can be regarded as processing energy. This distinction between information processing and energy processing serves to separate electronics from the rest of electrical engineering. Consequently, we view the nature of the discipline of electronics comprising the four C's-communication, computation, control, and components.

The prologue to this text provides a brief, coordinated historical review of modern electronics. The principal focus is on the development and applications of electronic devices and the growth of the industries resulting from the exploitation of these devices in practical circuits and systems.

Terminology

processing
information
electric energy

EXERCIS 11–

A. Words and Phrases

1. elusive
2. prologue
3. nuance
4. discipline
5. contributed to
6. flourishing
7. emerging
8. communicate with
9. province
10. conversion
11. distinction
12. principal
13. resulting from
14. exploitation

B. Reading Comprehension

1. Put T or F to indicate if the sentences below are true or false according to the facts in the passage.

- Charge motion confined to a metal is considered electronics. ________________
- Electrical engineers deal with systems by which we can communicate with one another worldwide. ________________
- The province of electrical engineering includes the devices used for the generation of thermal energy. ________________
- Information processing and energy processing are two major divisions of electrical engineering. ________________
- Electronics deals with energy processing. ________________

2. Answer the following questions.

- What does the four C's mean?
- What makes the distinction between electronics and the rest of electrical engineering?
- Define electronics in the strictest sense.

bibliography

Jacob Millman and Arvin Grabel, *Microelectronics*, pp. 1~2, McGraw-Hill, 1987.

4 Electromagnetism

We have all, at one time or another, enjoyed experimenting with a horseshoe or permanent bar magnet, as shown in Fig 11.1. When two north (or south) poles are brought close to each other, each makes every effort to avoid or repel the other. The reverse is true if opposite poles are brought together.

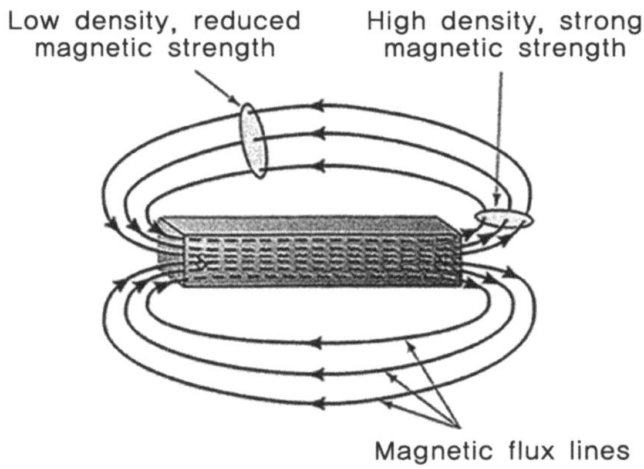

▲ Fig 11.1

Terminology
bar magnet
pole

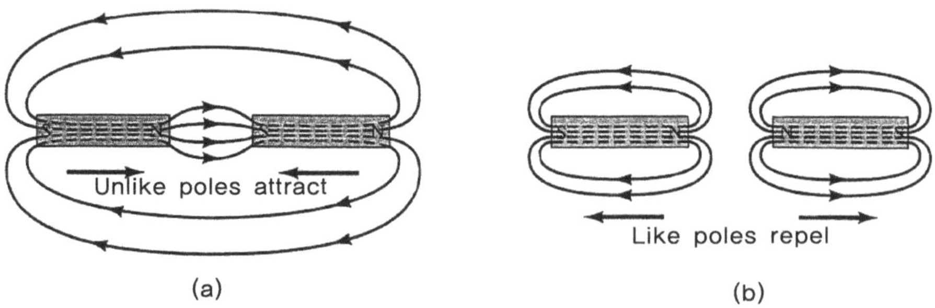

▲ Fig 11.2 (a) Attraction, (b) repulsion.

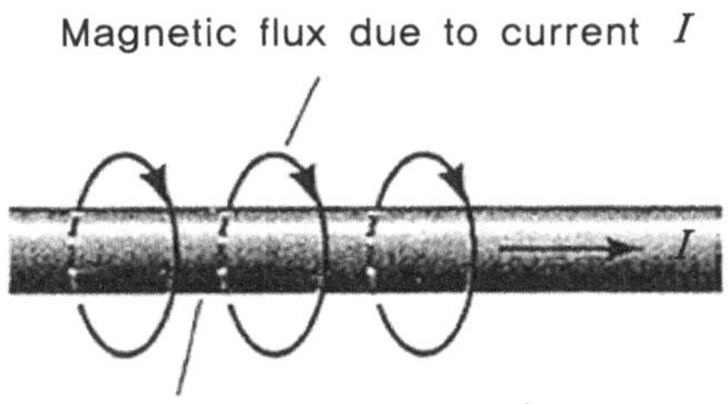

▲ **Fig 11.3** Magnetic flux lines surrounding a current carrying conductor.

In other words, like poles repel, while unlike poles attract. This phenomenon is demonstrated for each case in Fig 11.2 using magnetic flux lines, whose density can indicate the strength of the magnetic field in a particular region. They are continuous lines that leave the north pole of a magnet, return to the south pole, and pass through the magnetic material. Flux lines characteristically strive to be as short as possible in their continuous path. In Fig 11.2(a), therefore, the flux lines that pass directly from one magnet to the other tend to draw the two magnets together. For like poles a buffer action exists between the poles, as shown in Fig 11.2(b).

Permanent magnets such as appear in Fig 11.2 are constructed of a ferromagnetic material such as iron or steel that characteristically permits the setting up of flux lines through them with very little external pressure. Once magnetized, their retention properties are such that they exhibit magnetic properties after the external magnetizing agent is removed.

The current through a conductor establishes a circular flux pattern around the conductor such as shown in Fig 11.3. The direction of the flux lines can be determined by placing the thumb of the right hand in the direction of conventional current and noting the direction of the fingers. This determination is shown in Fig. 11.3. If we wrap a ferromagnetic material with a conduction wire as shown in Fig 11.4(a), a strong magnetic field can be established through the core.

Terminology

magnetic material
magnetic field
flux line
ferromagnetic
magnetic flux
conduction wire

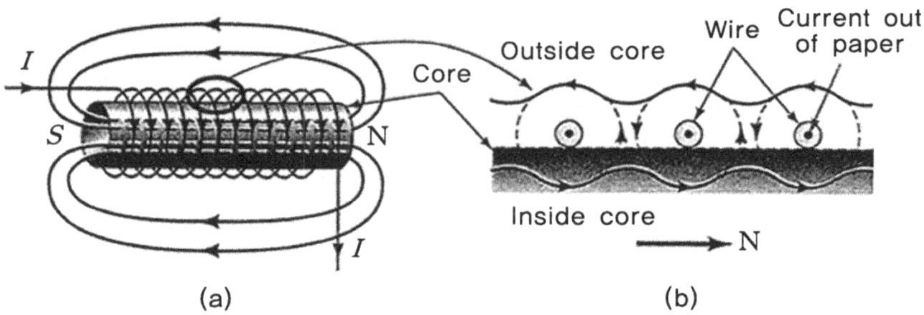

▲ **Fig 11.4** Electromagnet: (a) complete structure; (b) three adjacent windings on the top of the core.

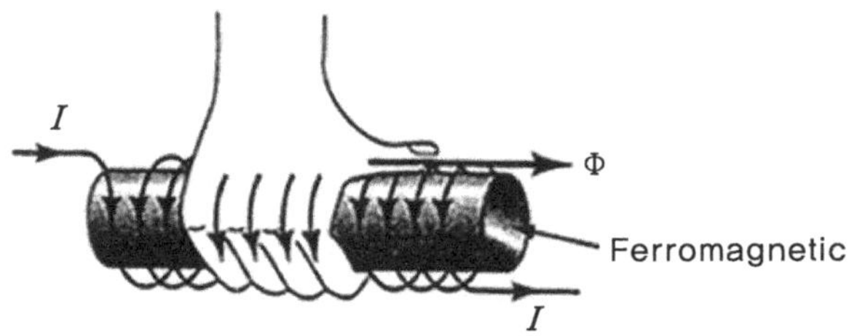

▲ **Fig. 11.5** Determining the direction of flux developed an electromagnet.

Terminology
windings
wrapped core
net flux

An expanded view of three adjacent windings on the top of the core is shown in Fig 11.4(b), with the resulting flux patterns for each loop as determined by the technique demonstrated in Fig 11.3. Note that between the windings there appears to be a canceling effect on the net flux, while above and below the two windings there is a strengthening of the flux in each direction. The net result for the wrapped core in Fig 11.4(a) is a north pole on the right, since the flux patterns are leaving, and a south pole on the left since the flux patterns are entering. The direction of the resultant flux can be determined by placing the fingers of the right hand in the direction of current for the wrappings, as shown in Fig 11.5 The thumb then points in the direction of the resultant flux. This particular structure, which requires electric current to develop its magnetic characteristics, is called an electromagnet. The strength of the magnet can be increased by increasing the current through the conductors. Eventually, however, a condition called saturation is reached such that any additional current has very little effect on the strength of the magnet. Once the magnetizing current is removed, the magnetic characteristics of the ferromagnetic core decreases to a value determined only by the ability of that sample to retain the magnetic characteristics — a measure of its retentivity. Materials with high retentivities are used to make permanent magnets by simply subjecting them to a strong magnctic field such as developed by the coil in Fig 11.4.

Terminology

electromagnet
retentivity
saturation

EXERCIS 11–

A. Words and Phrases

1. horseshoe
2. repel
3. reverse
4. phenomenon
5. strive to
6. retention
7. thumb
8. adjacent
9. canceling
10. effect on

B. Reading Comprehension

1. Put T or F to indicate if the sentences below are true or false according to the facts in the passage.

- When two opposite poles in a permanent magnet bar are brought together, they repel each other pole. ________________
- While unlike poles attract, like poles repel. ________________
- Permanent magnets can be constructed of ion. ________________
- Steel is a nonferromagnetic material. ________________
- A circular magnetic flux pattern can be generated around a conductor through which an electrical current flows. ________________

2. Answer the following questions.

- What is expected if two north poles are brought close to each other?
- According to the paragraph, what is the indication of the magnetic field strength in a particular region?
- Define a permanent magnet.
- What is electromagnet? Explain.

bibliography

Robert L. Boylestad and Louis Nashelsky, *Introduction to Electricity, Electronics, and Electromagnetics*, pp. 255~257, Prentice-Hall, 2002.

Unit 12

Digital and Analogue Systems

Unit 12 Digital and Analogue Systems

1 Digital and Analogue Systems

A **digital system** is a combination of devices designed to manipulate physical quantities or information that are represented in digital form; that is, they can take on only discrete values. These devices are most often electronic, but they can also be mechanical, magnetic, or pneumatic. Some of the more familiar digital systems include digital computers and calculators, digital watches, traffic-signal controllers, and typewriters.

An **analog system** contains devices that manipulate physical quantities that are represented in analog form. In an analog system, the quantities can vary over a continuous range of values. For example, the amplitude of the output signal to the speaker in a radio receiver can have any value between zero and its maximum limit. Other common analog systems are the telephone system, magnetic tape recording and playback equipment, and the automobile odometer.

Terminology

analog
digital
odometer
pneumatic

Advantages of Digital Techniques An increasing majority of applications in electronics, as well as in most other technologies, use digital techniques to perform operations that were once performed using analog methods. The chief reasons, for the shift to digital technology are:

1. *Digital systems are easier to design.* This is because the circuits that are used are *switching circuits*, where *exact* values of voltage or current are not important, only the range (HIGH or LOW) in which they fall.

2. *Information storage is easy.* This is accomplished by special switching circuits that can latch onto information and hold it for as long as necessary.

3. *Accuracy and precision are greater.* Digital systems can handle as many digits of precision as you need simply by adding more switching circuits. In analog systems, precision is usually limited to three or four digits because the values of voltage and current are directly dependent on the circuit component values.

4. *Operation can be programmed.* It is fairly easy to design digital systems whose operation is controlled by a set of stored instructions called a *program*. Analog systems can also be *programmed*, but the variety and complexity of the available operations is severely limited.

Terminology

IC chips
capacitor
inductor
switching circuits
transformer

5. *Digital circuits are less affected by noise.* Spurious fluctuations in voltage(noise) are not as critical in digital systems because the exact value of a voltage is not important, as long as the noise is not large enough to prevent us from distinguishing a HIGH from a LOW.

6. *More digital circuitry can be fabricated on IC chips.* It is true that analog circuitry has also benefited from the tremendous development of IC technology, but its relative complexity, and its use of devices that cannot be economically integrated (high-value capacitors, precision resistors, inductors, transformers), have prevented analog systems from achieving the same high degree of integration.

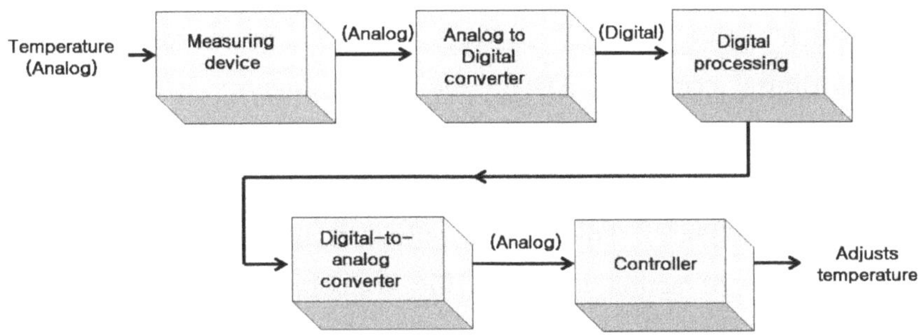

▲ **Fig 12.1** Block diagram of a temperature control system that requires analog/digital conversions in order to allow the use of digital processing techniques.

Limitations of Digital Techniques There is really only one major drawback when using digital techniques:

The real world is mainly analog.

Most physical quantities are analog in nature, and it is these quantities that are often the inputs and outputs that are being monitored, operated on, and controlled by a system. Some examples are temperature, pressure, position, velocity, liquid level, flow rate, and so on. We are in the habit of expressing these quantities *digitally*, such as when we say that the temperature is 64° (63.8° when we want to be more accurate); but we are really making a digital approximation to an inherently analog quantity.

To take advantage of digital techniques when dealing with analog inputs and outputs, three steps must be followed:

1. Convert the "real-world" analog inputs to digital form.
2. Process (operate on) the digital information.
3. Convert the digital outputs back to real-world analog form.

Fig 12.1 shows a block diagram of this for a typical temperature control system. As the diagram shows, the analog temperature is measured and the measured value is then converted to a digital quantity by an **analog-to-digital converter(ADC)**. The digital quantity is then processed by the digital circuitry, which may or may not include a digital computer. Its digital output is converted back to an analog quantity by a **digital-to analog converter(DAC)**. This analog output is fed to a controller which takes some kind of action to adjust the temperature.

The need for conversion between analog and digital forms of information can be considered a drawback because of the added complexity and expense. Another factor that is often important is the extra time required to perform these conversions. In many applications, these factors are outweighed by the numerous advantages of using digital techniques, and so the conversion between analog and digital quantities has become quite commonplace in the current technology.

There are situations, however, where using only analog techniques is simpler and more economical. For example, the process of signal amplification is most easily accomplished using analog circuitry.

It is becoming more and more common to see both digital and analog techniques employed within the same system in order to profit from the advantages of each. In these *hybrid* systems, one of the most important parts of the design phase involves determining what parts of the system are to be analog and what parts are to be digital.

Finally, it is safe to predict that as time goes on, we should see a greater and greater shift to digital techniques as the economic benefits of integration become even more overriding than they are now.

Terminology

hybrid system
DAC
digital computer
ADC
block diagram
signal amplification

EXERCIS 12–

A. Words and Phrases

1. manipulate	2. discrete	3. playback
4. complexity	5. spurious	6. fluctuation
7. prevent ~ from	8. inherently	9. take advantage of
10. conversion	11. drawback	12. commonplace
13. profit	14. integration	15. overriding

B. Reading Comprehension

1. Put T or F to indicate if the sentences below are true or false according to the facts in the passage.

- A digital system contains devices that manipulate physical quantities that are represented in digital form. ____________
- Analogue circuits are less affected by noise. ____________
- The precision is limited to five digits in analogue systems. ____________
- That the real world is mainly analogue is the only drawback of digital techniques. ____________
- Conversion between analogue and digital forms of information is expensive. ____________
- It is always simpler and more economical using digital techniques. ____________
- Operation can be easily programmed in analogue systems. ____________

2. Answer the following questions.

- What prevent analogue systems from achieving the same high degree of integration as digital systems?
- What is the reason that noise problem is not as critical in digital systems?
- What are the steps to take advantage of digital techniques when dealing with analogue input and output?
- What is DAC?
- Explain the hybrid system.

bibliography

Ronald J. Tocci, *Digital Systems: Principles and Applications*, pp. 4~6, Prentice-Hall, 1991.

2 Digital Computers(Heart of Automated Control)

In the following sections, we will examine how a computer works and what types of external computer components are required to monitor and control an automated process.

Computer Architecture

A computer consists of three types of devices, interconnccted by three sets of conductors called "buses." Fig 12.2 shows:

- central processor
- memory devices
- input/output devices

all interconnected via:

- a data bus
- an address bus
- a control bus

The **Central Processing Unit(CPU)** is the unit that controls the computer. The CPU reads the computer program from memory and executes the instructions. It manipulates values stored in memory and inputs or outputs data via input/output(I/O) devices. The CPU will be discussed in more detail later in this section.

The buses are sets of parallel conductors. An 8 bit computer has 8 conductors in its **data bus** and thus can move 8 bits of data at a time. A 16 (32, or 64) bit computer has 16 (32, or 64) data bus conductors and can process 16 (32, or 64) bit data "words" at a time. Some computers use bus-sharing techniques to reduce the number of conductors required

Terminology

instruction
central processor
control bus
I/O
CPU
address bus
data bus

in the data bus.

Each 8 bit memory location and I/O port has a unique address. The CPU uses the **address bus** to identify which of the memory locations or I/O locations it is about to read from or write to. Each conductor in the address bus can carry only either a 1 or a 0, so there is a limited number of different addresses that can be specified.

The Control bus contains miscellaneous other essential conductors. Some conductors provide the 5 volt supply and ground lines. Others are controllable by the computer components. The CPU, for example, uses the read/write line in the control bus to specify whether it intends to read from or write to the address that it has put on the address bus. There are several "interrupt" lines in the control bus that normally carry 5 volts and that, when dropped to 0 volts, cause the CPU to respond immediately even if a program is already running. Interrupts are used to initiate a computer respond to a situation that cannot wait, such as a fire in the warehouse, a computer reset, or simply a keyboard entry(which must be read before the human presses another key).

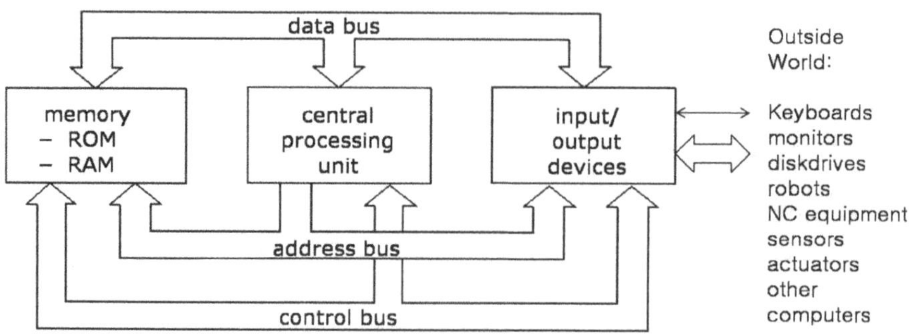

▲ Fig 12.2 Basic computer architecture

Some computer systems include a memory I/O line in the control bus, thereby as much as doubling the number of address that the computer can specify. Only half of the address can be in memory, while the other half must be I/O ports.

Memory in Fig 12.2 refers to **internal memory**. Programs on a disk or other mass storage device must be read into this memory via an I/O port before the CPU can run the program. There are basically two types of internal memory: random access and read only.

Random Access Memory(RAM) can be written to and read from by the **CPU**. RAM memory is lost when computer power goes off, unless a backup battery maintains power to the memory. Many industrial controllers (such as PLCs and robot and NC controllers) have battery-backed RAM, so that user-programs do not have to be reloaded into memory every time the controller is powered up.

Read Only Memory(ROM) can only be read by the CPU, and is not lost when the computer is turned off or unplugged. ROM chips come in several types. Some ROM chips are **custom manufactured** where large volumes warrant. **PROM** chips can be programmed once by the user, using a programming device. They are used by medium-quantity suppliers if technology changes fast enough to rule out the cost of tooling-up for custom ROM(e.g.,a robot supplier). Erasable PROMs(**EPROM**) can be programmed like PROMs, but can be erased for reuse by exposure to high-intensity ultraviolet light. They are used by computer suppliers who write their own programs for storage in ROM, but who make frequent changes to those programs.

Electrically Erasable PROMs(**EEPROM or EAPROM**) are commonly supplied with PLCs. They are used very much like floppy disks, but they do not require expensive, fragile read/write mechanisms. User-programs can be copied from PLC RAM onto an EEPROM chip in a plastic carrier, then the EEPROM can be removed for use at another PLC, or can be left

Terminology

ultraviolet
floppy disk
RAM
random access
PLC
EPROM
PROM
ROM
EEPROM
internal memory

in place for copying back into RAM at startup.

Flash memory, often on removable "**PCMCIA**" (Personal Computer Memory Card International Association) cards, is now available. The computer reads and writes flash memory as if was RAM, but data is not lost when power is removed.

Terminology

Flash memory

EXERCIS 12–

A. Words and Phrases

1. execute	2. in more detail	3. miscellaneous
4. interrupt	5. warehouse	6. go off
7. turned off	8. unplugged	9. warrant

B. Reading Comprehension

1. Put T or F to indicate if the sentences below are true or false according to the facts in the passage.

- A computer is composed of three types of devices which are central processor, memory devices and input/output devices. ____________
- The address bus is used to identify which of the memory locations or I/O Locations it is about to read from or write to. ____________
- PCMCIA is the abbreviation of 'Personal Computer Memory Center for International Academy'. ____________
- ROM can only be read by CPU. ____________
- RAM can be written by CPU. ____________
- PROM chips can be programmed several times by the user. ____________
- All the control buses provide the 5 volt supply. ____________
- Data in the flash memory is not lost when power is removed. ____________

2. Answer the following questions.

- What kinds of internal memories are there in a computer? List them all.
- What are the advantages of EEPROM? List them all.
- What is the function of the CPU?
- How many buses are used to interconnect the computer devices? List them all.

bibliography

S. Brian Morriss, *Automated Manufacturing Systems*, pp. 197~199, McGraw-Hill, 1994.

Unit

13

Internal Combustion Engine and Measurement

Unit 13 Internal Combustion Engine and Measurement Systems

1 Internal Combustion Engine

The internal combustion engine is a heat engine in which the burning of a fuel occurs in a confined space called a combustion chamber. This exothermic reaction of a fuel with an oxidizer creates gases of high temperature and pressure, which are permitted to expand. The defining feature of an internal combustion engine is that useful work is performed by the expanding hot gases acting directly to cause movement, for example by acting on pistons, rotors, or even by pressing on and moving the entire engine itself.

This contrasts with external combustion engines such as steam engines which use the combustion process to heat a separate working fluid, typically water or steam, which then in turn does work, for example by pressing on a steam actuated piston.

Terminology

combustion chamber
external combustion
oxidizer
crankshaft
engine
exothermic reaction
four strokes
heat engine
steam engine

The basic four-stroke cycle diesel engine

From the name, it is fairly obvious that there are four strokes in one complete engine cycle. A stroke is the movement of the piston through the full length of the cylinder, and, since one such movement causes the crankshaft to rotate half a turn, it follows that there are two crankshaft revolutions in one complete engine cycle. The four strokes, in correct order, are:

(i) The inlet stroke. With the inlet valve open and the exhaust valve closed, the piston moves from TDC to BDC, creating a low-pressure area in the cylinder. Clean, filtered air rushes through the open inlet valve to relieve this low-pressure area, and the cylinder fills with air.

(ii) The compression stroke. With both valves closed, the piston moves from BDC to TDC, compressing the air. During this stroke the air becomes heated to a temperature sufficiently high to ignite the fuel.

(iii) The power stroke. At approximately TDC, the fuel is injected, or sprayed, into the hot, compressed air, where it ignites, burns and expands. Both valves remain closed, and the pressure acts on the piston crown, forcing it down the cylinder from TDC to BDC.

(iv) The exhaust stroke. At approximately BDC the exhaust valve opens and the piston starts to move from BDC to TDC, driving the burnt gas from the cylinder through the open exhaust valve.

Terminology

compression stroke
inlet stroke
power stroke
flywheel
BDC
TDC
inlet valve
cylinder
exhaust stroke
exhaust valve
single–cylinder

At the completion of the exhaust stroke, the exhaust valve closes, the inlet valve opens and the piston moves down the cylinder on the next inlet stroke. Since there are three non-working strokes to one working stroke, some means of keeping the engine turning over must be provided, particularly in single-cylinder engines. It is for this reason, and to ensure smooth running, that a heavy flywheel is fitted to the crankshaft.

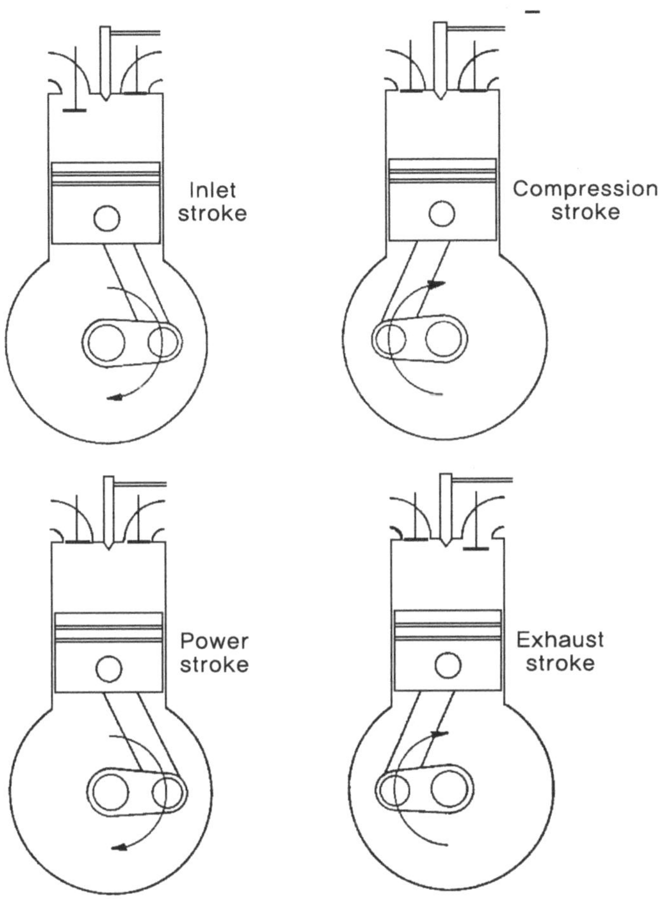

▲ Figure 13.1

Terminology
scavenging

Scavenging the four-stroke cycle diesel engine

It is necessary, since efficient combustion is desired, to completely clear the burnt gas from the cylinder on the exhaust stroke, so providing a full cylinder of fresh air by the completion of the inlet stroke. Air which has burned with fuel has had almost all its oxygen consumed, and cannot be used again. The clearing of the exhaust gas from the cylinder of an internal combustion engine is known as 'scavenging'.

Both the upward movement of the piston on the exhaust stroke and the valve timing contribute to the scavenging of the burnt gas in the four-stroke engine.

It is usual for the exhaust valve to be opened before BDC, thus allowing a puff of high-pressure gas to escape at high velocity through the exhaust system. The upward movement of the piston ensures that the burnt gas continues this high-speed movement.

At the end of the exhaust stroke, the gas will continue to move through the exhaust system because of its momentum alone. This continued movement will cause a low-pressure area to develop behind the fast-moving exhaust gas, and this can be used to draw fresh air into the cylinder if the inlet valve opens at the instant the low-pressure area develops.

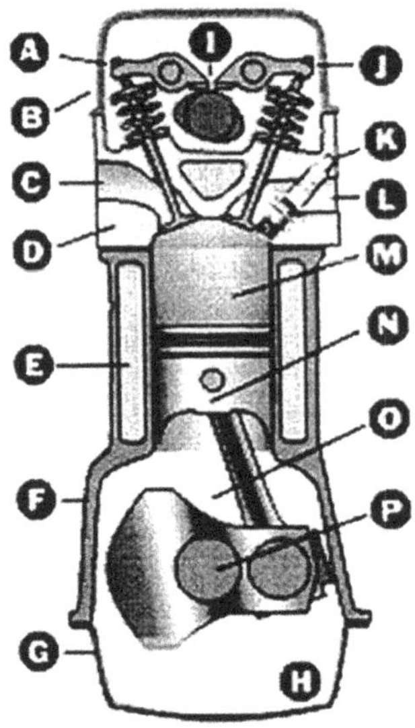

- A Intake Valve. Rocker Arm & Spring
- B Valve Cover
- C Intake port
- D Head
- E Coolant
- F Engine Block
- G Oil Pan
- H Oil Sump
- I Camshaft
- J Exhaust Valve. Rocker Arm & Spring
- K Spark Plug
- L Exhaust Port
- M Piston
- N Connecting Rod
- O Rod Bearing
- P Crankshaft

▲ Fig 13.2 Cross-section of internal combustion engine

In practice, it is found that good scavenging can be achieved by opening the inlet valve just before TDC, and closing the exhaust valve just after TDC, thus allowing the exhaust gas to 'draw' fresh air into the cylinder before the inlet stroke actually commences.

The period of crankshaft rotation when both valves are open together is known as **valve overlap**, and occurs only at TDC.

Suitable positioning of the valves, usually one at each side of the combustion chamber, helps to guide the gases on their correct paths, and assists in achieving complete scavenging.

It is worthwhile to note at this stage that supercharged engines usually have more valve overlap than naturally aspirated engines.

Terminology

supercharged engine
valve overlap
naturally aspirated engine
combustion chamber

EXERCIS 13–

A. Words and Phrases

1. confine
2. expand
3. contrast with
4. revolution
5. rush through
6. spray
7. completion
8. consume
9. clear
10. ensure
11. momentum
12. commence
13. worthwhile
14. overlap
15. exothermic
16. oxidiser
17. aspirate

B. Reading Comprehension

1. Put T or F to indicate if the sentences below are true or false according to the facts in the passage.

- There are four strokes in one complete cycle in the basic four-stroke cycle diesel engine. ________________
- Both the inlet and exhaust valves remain open during the power stroke. ________________
- Scavenging is a clearing process of the exhaust gas from the cylinder of an internal combustion engine. ________________
- Good scavenging can be achieved by opening the inlet valve just before TDC. ________________
- Supercharged engines usually have less valve overlap than naturally aspirated engines. ________________
- The inlet valve closes at the completion of the exhaust stroke. ________________

- It is the upward movement of the piston on the exhaust stroke that contribute to the scavenging of the burnt gas in the four–stroke engine. ____________

- Suitable positioning of the valves is very important in achieving complete scavenging. ____________

2. Answer the following questions.

- Please define an internal combustion engine according to the paragraph.
- How do the valves act during the inlet stroke?
- Why is the scavenging process necessary in internal combustion engines?
- What kinds of valve actions are desired for good scavenging?
- How many non–working strokes are there in a four–stroke engines?

bibliography

A.F. Asmus and S.C. Gough, *Diesel Engines and Fuel Systems*, pp. 11~12, Pitman Publishing Pty Ltd, 1979.

2 Measurement System

A measurement is an act of assigning a specific value to a physical variable. That physical variable becomes the measured variable. A measurement system is a tool used for this quantification of the physical variable.

As such, it is used to extend the abilities of the human senses that, while they can detect and recognize different degrees of roughness, length, sound, color, and smell, are limited and are not very adept at assigning specific values to sensed variables. A general template for a measurement system is illustrated in Figure 13.3. Basically such a system consists of part or all of four general stages:

1. Sensor-transducer stage
2. Signal-conditioning stage
3. Output stage
4. Feedback-control stage

Terminology

thermal equilibrium
physical variable
quantification
measurement
calibration
measured variable
Feedback−control
Sensor−transducer
Signal−conditioning

These stages form the bridge between the input to the measurement system and the system output, a quantity which is used to infer the value of the physical variable measured. The relationship between the input information, as acquired by the sensor, and the system output is established by a calibration.

The sensor is a physical element that employs some natural phenomenon by which it senses the variable being measured. The transducer converts this sensed information into a detectable signal form, which might be electrical, mechanical, optical, or otherwise. The goal is to convert the sensed information into a form that can be easily quantified. For example, the liquid contained within the bulb on the common bulb thermometer of Fig 13.2 exchanges energy with its surroundings until the two are in thermal equilibrium. At that point they are at the same temperature. The phenomenon of thermal expansion of the liquid results in its movement up and down the stem. The liquid in the bulb acts as the sensor. By forcing the expanding liquid into a narrow capillary, this measurement system transforms thermal information into a mechanical displacement.

Hence, the bulb's internal design acts as a transducer.

It is worth noting that the term "transducer" is also often used in reference to a packaged device, which may contain a sensor, transducer, and some signal conditioning elements. The context in which the term is used usually prevents any ambiguity.

Sensor selection, placement, and installation are particularly important, because the input to the measurement system is the information sensed by the sensor. Accordingly, the interpretation of all information passed through and indicated by the system depends on that which is actually sensed by the sensor.

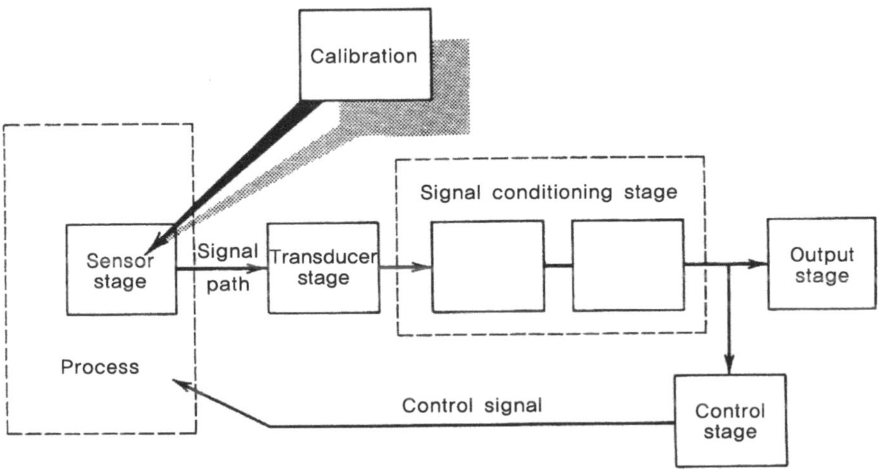

▲ **Figure 13.3** Components of a general measurement system.

Terminology

capillary

Signal conditioning equipment takes the transducer signal and modifies it to a desired magnitude. This optional intermediate stage might be used to perform tasks such as increasing the magnitude of the signal through amplification, removing portions of the signal through some filtering technique, and/or providing mechanical or optical linkage between the transducer and the output stage, for example, converting a translational displacement of a sensor into a rotational displacement of a pointer. This stage can consist of one or more devices, which are often connected in series. For example, the diameter of the thermometer capillary determines how far up the stem the liquid moves with increasing temperature. It "conditions" the signal.

The output stage provides an indication of the value of the measurement. The output equipment might be a simple readout display, a marked scale, or might contain devices that can record the signal for later analysis. Examples of these devices are tape recorders, chart recorders, and computer disk drives. The scale of the bulb thermometer in Fig 13.4 serves as the output stage of that measurement system.

Terminology

translational
furnace thermostat
rotational
thermometer

In those measurement systems involved in process control, a fourth stage, the feedback-control stage, contains a controller that interprets the measured signal and makes a decision regarding the control of the process. This decision results in a change in a process parameter that affects the magnitude of the sensed variable. In simple controllers, this decision is based on the magnitude of the signal of the sensed variable, usually whether it exceeds some high or low set point, a value set by the system operator. A simple measurement system with control stage is a household furnace thermostat. The operator fixes the set point for temperature on the thermostat display, and the furnace is activated as the local temperature at the thermostat, as determined by the senor within the device, rises or falls about the set point. In a more sophisticated controller, a signal from a measurement system can be used as an input to an "expert system" controller that, through an artificial intelligence scheme, determines the optimum set conditions for the process.

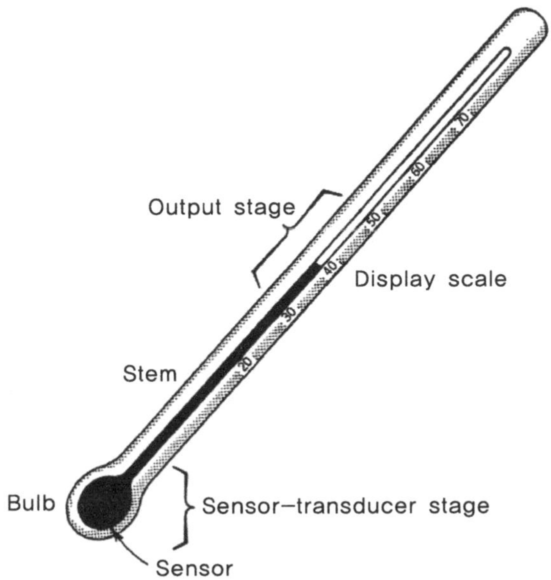

▲ **Figure 13.4** Components of bulb thermometer equivalent to sensor, transducer, and output stages.

Terminology

artificial intelligence

EXERCIS 13–

A. Words and Phrases

1. roughness
2. stem
3. template
4. quantity
5. infer
6. be acquired by
7. employ
8. phenomenon
9. detectable
10. quantified
11. result in
12. ambiguity
13. portion
14. filter
15. readout
16. interpret
17. exceed
18. sophisticated
19. scheme
20. optimum
21. context
22. expert

B. Reading Comprehension

1. Put T or F to indicate if the sentences below are true or false according to the facts in the passage.

- The transducer converts the sensed information into a detectable signal form. ________
- Signal conditioning equipment simply takes the transducer signal and transfers it to the display panel. ________
- Physical variables cannot be converted into the measured variables. ________
- The sensor is a physical element that employs specific values. ________
- In simple controllers, the decision is based on the magnitude of the signal. ________
- The goal of the transducer is to convert conditioned informations into physical variables. ________

- The bulb thermometer utilizes the principle of thermal equilibrium.

  ____

- The liquid in the bulb of a bulb thermometer acts as a sensor.

  ____

2. Answer the following questions.

- What is the function of the signal conditioning equipment?
- List four stages of general measurements.
- What is the definition of a measurement?
- What are the functions of a controller in the feedback-control stage?
- What are particularly important in sensing the correct input to the measurement system?

bibliography

Richard S. Figliola and Donald E. Beasley, *Theory and Design for Mechanical Measurements*, pp. 2~4, John Wiley & Sons, 1995.

Unit

14

Mechatronics and Controls

Unit 14 Mechatronics and Controls

1 Mechatronics

Mechanical systems have become increasingly dependent on computers and electronics to achieve the degree of function, flexibility, and reliability demanded by users. The combination of mechanics and electronics has been called mechatronics, a term that has come to include computers as well. A mechanical system can be characterized as

system = mechanics + electronics + software

The mechanics includes the mechanisms, electromagnetic components, thermal components, flow elements, and so on. The most notable function of the system is usually realized through the mechanics section. The instruments and actuators are considered part of the system mechanics.

The overall control of the mechanical system is expressed in the software. It is the unique role of software that differentiates modern mechanical systems from their forebears. Control of mechanical system through software affords a degree of flexibility unknown to the traditional machine designer. Changing, for example, from manufactured parts of one dimension to parts with another dimension is only a matter of changing numbers in a program. In traditional machines it might have required changing gears, resetting linkages, or other mechanical operations that could take many

Terminology

gear
electronics
actuator
electromagnetic
mechanics
linkage
mechatronics

hours improving its reliability and probably decreasing its cost.

With this perspective, a new definition for mechatronics can be formulated:

Mechatronics is the application of complex decision making to the operation of physical systems.

This definition recognizes the central role played by software and computers, the decision makers in the design and implementation of mechanical systems. It also recognizes the possibilities of using other technologies for decision making in the future.

Electronics forms the glue that connects the mechanics to the software. At the time the term mechatronics was coined, by combining mechanics and electronics, the electronics components themselves were so much more flexible and easy to program than mechanical components that a new class of machine was defined. Software has, however, eclipsed electronics in flexibility and decision making, defining yet another level of intelligent machines. Electronics continues to play a crucial role. That is the role of interface between mechanical components and the computers that run the system software.

Terminology

intelligent machine
implementation

EXERCIS 14–

A. Words and Phrases

1. increasingly
2. flexibility
3. reliability
4. notable
5. differentiate
6. forebear
7. perspective
8. glue
9. eclipsed
10. interface

B. Reading Comprehension

1. Put T or F to indicate if the sentences below are true or false according to the facts in the passage.

- Control of mechanical system through software has disadvantages in flexibility. ____________
- Electronics has the role of interface between mechanical components and the computers that run the system software. ____________
- Electronics has more flexibility than software. ____________
- Software is indispensible for intelligent machines. ____________
- Electromagnetic components are included in the mechanics. ____________
- The overall control of the mechanical system can be accomplished by electronics only. ____________

2. Answer the following questions.

- Please name three fields of which mechanical systems consist.
- How is the name 'mechatronics' made?
- According to the paragraph what is a new definition for mechatronics?

bibliography

David M. Auslander and Carl J. Kempf, *Mechatronics Mechanical System Interfacing*, pp. 1~2, Prentice-Hall, 1996.

2 Feedback Control Systems

Various forms of transportation are illustrated in Fig 14.1. All the basic concepts of feedback control systems are contained in each of these means of transportation. For the basic form of transportation, walking, the desired speed at which the walker wishes to go is the reference input. When the walker takes a leisurely stroll through a park, the desired speed is slow. When the person is in hurry, desired speed is fast. The actual speed is the controlled variable (i.e., the quantity being controlled). The part of a system which compares the reference input (desired value) with the controlled variable (actual value) in order to measure the error is called the comparator. The brain serves as the comparator for the walker. Typically, the error signal goes to a power-amplifying device (muscles in the legs) which actuates the system to be controlled so as to reduce the error to zero. Thus, the actual speed is the same as the desired speed.

Terminology

reference input
comparator
feedback control
transportation
controlled variable

▲ Fig 14.1 Various forms of transportation.

The driver of a car compares the actual speed with the desired speed. Again the brain serves as the comparator. In response to the error signal, the driver changes the position of the accelerator pedal. The engine serves as the power-amplifying device. That is, a little motion of the accelerator pedal causes a substantial change in the power produced by the engine.

For cruise control, a magnet is attached to the driveshaft which rotates at some constant times the speed of the wheels. A magnetic pickup fixed to the body of the car generates a pulse every time the driveshaft makes one complete rotation. The rate at which pulses are being generated is a measure of the actual speed of the car. At the instant the cruise control is set, the rate at which pulses are being generated is stored so as to serve as the reference input (i.e., desired speed). A vacuum-operated device is attached to accelerator linkage. When the actual speed is greater than the reference speed, the rate at which pulses are being generated is greater than the reference value. This error signal goes to the vacuum device which serves as a power amplifier to change the position of the accelerator linkage so as to decrease the actual speed.

The most difficult problem which the Wright brothers encountered in the process of making the airplane into a useful machine was the control problem. After much experimentation, they developed an ingenious pulley system coordinating the motion of the aileron and the rudder so as to produce a smooth turning effort. Note that the pilot (Orville or Wilbur) would compare the actual path of the plane with the desired path, and then in response to the error signal he would actuate the pulley system. Aerodynamic forces acted on the wings and rudder would serve as the power-amplifying device to change the trajectory of the plane. Although modern airplanes have much more sophisticated means for controlling the position of the ailerons and rudder, the principles are still the same.

The controlling of temperature is a typical of a feedback control system. The position of the temperature dial sets the desired temperature (i.e., the

Terminology

power amplifier
driveshaft
aileron
aerodynamic forces
vacuum device
rudder
cruise control
magnetic pickup
accelerator pedal

reference input). The actual temperature of the system is the controlled variable (i.e., the quantity which is being controlled). The thermostat, or comparator, compares the actual temperature with the desired temperature in order to measure the error. This error signal is the actuating signal, which is then sent to the heating units in order to correct the temperature. For example, if the actual temperature is less than desired temperature, the actuating signal causes the heating elements to supply more heat. If there is no error, the actuating signal does not change the amount of heat which is being supplied. When the actual temperature is greater than the desired value, then the actuating signal calls for decrease in the amount of heat.

For a system to be classified as a feedback control system, it is necessary that the controlled variable be fed back and compared with the reference input. In addition, the resulting error signal must actuate the control elements (power amplifying device) to change the output so as to minimize the error. A feedback control system is also called a *closed-loop system*. Any system which incorporates a thermostat to control temperature is a feedback, or closed-loop, system. Well-known examples are electric frying pans, irons, refrigerators, and household furnaces with thermostatic control.

For speed control systems, the device which subtracts the feedback signal from the reference input (i.e., the comparator) is often a centrifugal governor. The compression of a spring sets a force which is a measure of the desired speed. The centrifugal force of the flyweights is a measure of the actual speed. The difference in these forces is a measure of the error. The governor serves the same purpose that the thermostat does for temperature controls. That is, the governor compares the actual speed with desired value and measures the error. This error signal then goes to a power amplifier such as a hydraulic servomotor that controls the position of a flow valve which in turn determines the rate of fuel flow to the engine. The same basic concepts apply to all types of feedback control systems, whether the controlled variable is temperature, speed, pressure, flow,

Terminology

closed−loop system
power amplifier
hydraulic servomotor
thermostat
centrifugal governor
centrifugal force

position, force, torque, or any other physical quantity.

In an open-loop system, there is no comparison of the controlled variable with the desired input. Each setting of the input determines a fixed operating position for the control elements. For example, for a given input temperature setting, the heating units are positioned to supply heat at a fixed rate. Note that there is no comparator, or thermostat, which measures the error and resets the heating units. The disadvantage of such a system is illustrated by the fact that if a fixed rate of heat is supplied to a house, the inside temperature will vary appreciably with changes in the outside temperature. Thus, for a given set input to an open-loop system, there may be a big variation of controlled variable depending on the ambient temperature.

In this example, the ambient temperature is an external disturbance. By an external disturbance is meant something external to the system which acts to change or disturb the controlled variable. A major advantage of employing feedback control is that, because of the comparator, the actuating signal continually changes so that the controlled variable tends to become equal to the reference input regardless of the external disturbance. Another consideration is that with feedback one can generally use relatively inexpensive components and yet obtain better control than is possible with very expensive components in an open-loop system. The primary focus of this text is feedback control systems.

Terminology

open−loop system
external disturbance

EXERCIS 14–

A. Words and Phrases

1. leisurely
2. stroll
3. in hurry
4. actuate
5. substantial
6. ingenious
7. sophisticated
8. principle
9. compression
10. comparison
11. disadvantage
12. appreciably
13. variation
14. ambient
15. disturb
16. regardless of

B. Reading Comprehension

1. Put T or F to indicate if the sentences below are true or false according to the facts in the passage.

- A feedback control system is more expensive than an open–loop system for the same effect of control. ________________
- One can use a centrifugal governor for speed control systems. ________________
- There is no comparison of the controlled variable with the desired input in a closed-loop system. ________________
- A feedback control system is also called a closed-loop system. ________________
- Control problem was the most difficult problem which the Wright brothers encountered in the process of making the airplane. ________________
- When a man drives a car, the brain serves as the comparator. ________________
- The controlling of temperature is a typical of an open–loop system. ________________

- The pulley system developed by the Wright brothers is composed of a rudder, a thermostat and some magnetic pickups. ____________
- In an open–loop system for controlling room temperature, the ambient temperature is an external disturbance. ____________
- There always exists comparison of the controlled variable with the desired input in an open–loop system. ____________

2. Answer the following questions.

- What is necessary for a system to be classified as a feedback control system?
- What would serve as the power-amplifying device to change the trajectory of the plane?
- What is the definition of the comparator in a feedback control system?
- What do you call the difference between the reference input and the controlled variable in a feedback control system?
- For speed control systems, what has the same purpose as the thermostat for temperature controls?
- Please define an external disturbance according to the paragraph.
- What are the advantages of employing feedback control instead of open-loop system?

bibliography

Francis H. Raven, *Automatic Control Engineering*, pp. 3~5, McGraw–Hill, 1995.

Unit 15

Industrial Robot

Unit 15 Industrial Robot

1 Industrial Robot

General Components of an industrial robot

An industrial robot has three types of components: physical parts or anatomy, built-in instructions or instinct(placed there by the manufacturer), and learned behavior or task programs(on-the-job training).

The physical portion of an industrial robot is made up of three or four parts: the mechanical part or manipulator, which performs the work through motion; the controller, which directs motions; the power supply, which supplies energy to the manipulator; and (optionally) the vehicle, which transports the base of the robot to where it is to do its work. Fig 15.1 draws comparisons between the physical parts of the human being and those of the industrial robot.

The **manipulator** is the part of the robot that physically performs the task. Attached to it is the gripper or hand, which actually contacts the parts or materials being processed.

The **controller** is the brain or computer center of the robot. It may be as simple as a set of mechanical stops and limit switches or as complex as a complete minicomputer or microcomputer. The controller must recall all the motions that the robot is to perform, in the proper order and with the correct timing. When two or more robots are designed to work together, their individual controllers may receive directions from a large electronic

Terminology

gripper
industrial robot
minicomputer
manipulator
power supply

computer that controls the entire factory. Plotting the movements of several robots working together is known as **robot choreography**.

The **power supply** for the manipulator depends on whether the type of power used by the manipulator is electric, hydraulic, pneumatic, or vacuum.

The robot may be stationary or sit on a vehicle. The vehicle may use wheels on a rack, free wheels, treads, or some type of legs and feet. Whether the robot is stationary or mobile, the base of the robot must be stable and accurately situated if the robot's manipulator is to perform accurately.

Terminology

pneumatic
tread
hydraulic
rack
choreography

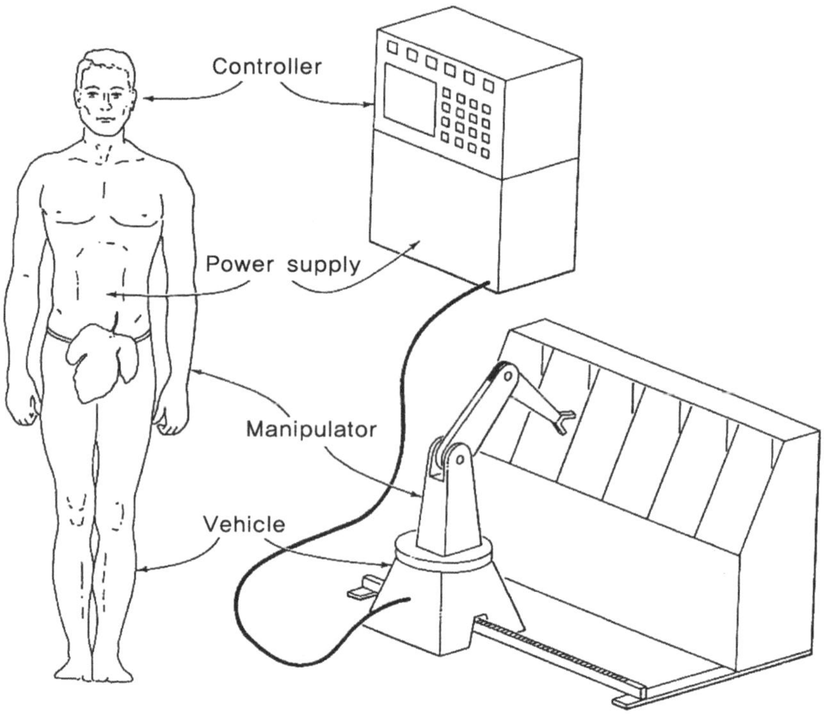

▲ Fig 15.1 parts of a Robot

The manipulator, controller, and power supply for the robot may be housed together as a single unit or they may be housed separately as discrete units. To apply spray paint, for example, only the manipulator need actually be located in the spray-paint booth. The controller and power supply can be kept in another area with a less severe environment. If a robot works in an explosive hazard area, again only the manipulator need be present in the area.

The controller for the human is the brain; the controller for the industrial robot is an electronic computer. The power supply for the human is the digestive system; for the industrial robot, it can be electrical (as in this figure), pneumatic, or hydraulic. The main manipulator for the human is the arm, while for the robot it is the manipulator arm. The vehicle for the human is the legs; for the industrial robot in this figure, it is wheels that move along a fixed rail.

Manipulator Configurations

Without the manipulator, the robot would not be a robot since it would not be able to produce motion. The parts of a robot's manipulator are named after similar parts in a human's chief manipulator, the arm and the hand. The points at which a robot's manipulator bends, slides, or rotates are called joints or axes of motion; they include the shoulder, the elbow, the wrist, and the finger joints. The number of joints an industrial robot has determines its degrees of freedom of motion. Each axis equals one degree of freedom.

Fig 15.2 draws comparisons between human and robot manipulators. The human arm and hand constitute the finest universal manipulator in nature. Thus they serve as a model for the industrial robot to copy. The industrial robot of today generally has only a single manipulator, with one arm and one hand on it. In a few cases, the arm may have more than one hand on it. The method by which the arm positions the hand varies from one

Terminology

wrist
joint
degrees of freedom

robot to another.

Robots are often classified by the shape of the space or **work envelope** that their manipulator can reach. For example, the manipulator of a **rectangular**—or **Cartesian-coordinates robot** operates within a rectangular-shaped work envelope.

The industrial robot manipulator has an arm, a wrist, and a hand with fingers. These names match those of the corresponding human parts. The jointed-arm manipulator shown here also has an upper arm and a forearm.

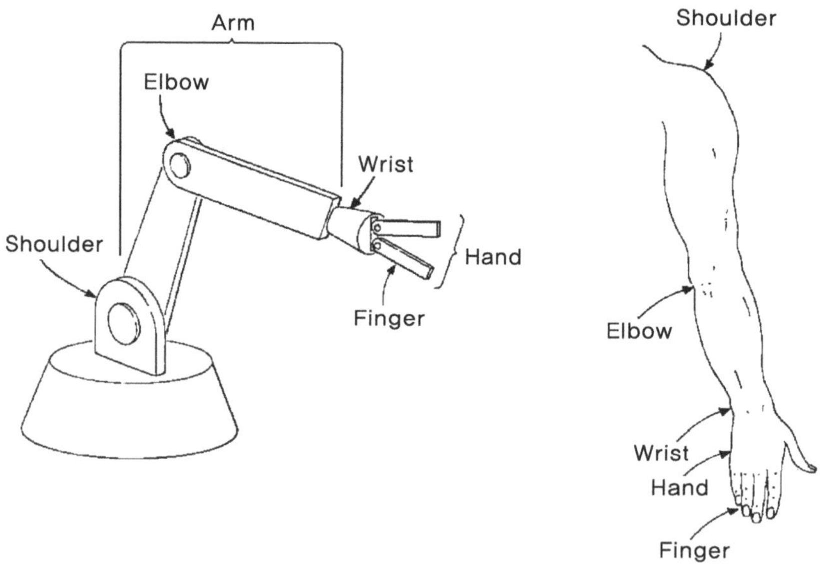

▲ **Figure 15.2** Parts of a Manipulator.

Terminology

work envelope
Cartesian–coordinates
rectangular–coordinates

EXERCIS 15–

A. Words and Phrases

1. anatomy
2. be made up of
3. stationary
4. explosive
5. hazard
6. discrete
7. digestive
8. constitute
9. forearm
10. name after

B. Reading Comprehension

1. Put T or F to indicate if the sentences below are true or false according to the facts in the passage.

- The manipulator is the part of a robot which supplies the power. ________
- When two or more robots work together in the factory, they are controlled by a central electronic computer in general. ________
- The controller for the industrial robot is the electronic computer. ________
- The industrial robot manipulator does not have the names such as an arm, a wrist and a hand. ________
- Degree of freedom of motion of a robot is determined by the number of power supplies. ________
- A robot cannot produce motion with the manipulator. ________
- The manipulator, the controller and the power supply must be housed together in any circumstances. ________
- An industrial robot must have a vehicle on which it can sit. ________

2. Answer the following questions.

- What are three important parts of a robot?
- After what are the parts of a robot's manipulator named?
- What kinds of method are used for the movement of the robot vehicle?
- What is the function of the manipulator of a robot?
- Explain the term 'robot choreography'.

bibliography

James L. Fuller, *Robotics: Introduction, Programming, and Projects*, pp. 38~40, MacMillan Publishing Company, 1991.

2) Factory Automation

Factory automation is one of the major objectives of modern industry. Although there is no one plan for factory automation, a general configuration is presented in Fig 15.3. In this configuration, all branches of the factory communicate and exchange information through a central data base. Various parts of the product are designed in the computer-aided engineering (CAE) branch, and then the design is sent to the computer-aided manufacturing (CAM) branch for parts manufacturing and final assembly. Two of the major subbranches of CAE are computer-aided product design and computer-aided manufacturing design.

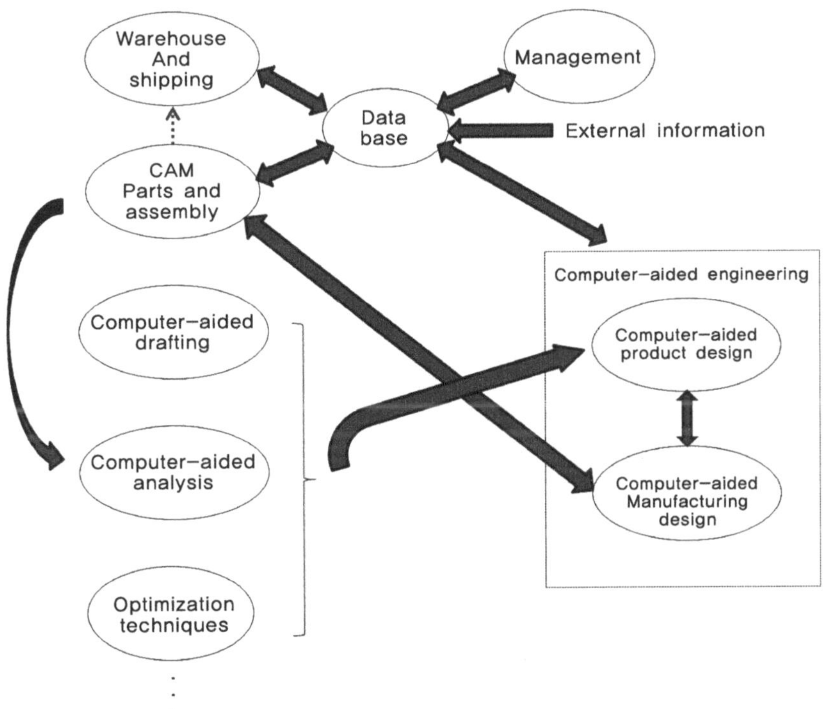

▲ **Fig 15.3** Automated design and manufacturing

Terminology

CAM
CAE
data base
factory automation

The computer-aided product design branch, better known as computer-aided design (CAD), may consider the design of single parts or it may concern itself with the final product as an assembly of those parts. Computerized product design requires such capabilities as computer-aided analysis, computer-aided drafting, design sensitivity analysis, or optimization. The computer-aided analysis capability serves as part of the design process and is also used as a model simulator for the finished manufactured product. Analysis may be considered especially appropriate for a product whose initial design has to be modified several times during the manufacturing process. Thus computer-aided analysis can be used as a substitute for laboratory or field tests in order to reduce the cost.

The computer-aided manufacturing design branch is concerned with the design of the manufacturing process. This branch considers the manufacturability of newly designed parts and employs techniques to improve the manufacturing process, in addition to on-line control of the manufacturing process.

Terminology

manufacturing process
on−line control
CAD
manufacturability
drafting
sensitivity analysis
simulator
optimization

EXERCIS 15−

A. Words and Phrases

1. objective
2. configuration
3. subbranch
4. capability
5. appropriate
6. substitute
7. drafting
8. simulator
9. manufacturability

B. Reading Comprehension

1. Put T or F to indicate if the sentences below are true or false according to the facts in the passage.

- Various parts of the product are designed in the computer−aided manufacturing branch. ________
- Computer−aided analysis can be used as a substitute for factory manufacturing. ________
- The computer−aided manufacturing design considers the manufacturability of newly designed parts. ________
- In a good factory automation, all branches communicate and exchange information through a central data base. ________
- CAM is the abbreviation of computer−aided machining. ________
- The computer−aided product design is known as computer−aided design. ________
- The computer−aided analysis capability can also be used as a model simulator for the finished manufactured product. ________

2. Answer the following questions.

- What are two of the major subbranches of CAE?
- What capabilities are required in the computerized product design?
- What kind of method should be used when the initial design has to be modified several times during the manufacturing process?
- What does the computer–aided manufacturing design branch consider in the design of the manufacturing process?
- List some capabilities which the computerized product design requires.

bibliography

Parviz E. Nikravesh, *Computer–Aided Analysis of Mechanical Systems*, pp. 1~2, Prentice–Hall, 1988.

3) Reliability, Quality, and Safety

Reliability is increasingly becoming important during the design of engineering systems, as our daily lives and schedules are more dependent than ever before on the satisfactory functioning of these systems. Some examples of these systems are computers, trains, automobiles, aircraft, and space satellites.

Some of the specific factors that are playing a key role in increasing the importance of reliability in designed systems include system complexity and sophistication, competition, increasing number of reliability/ safety/ quality-related lawsuits, public pressures, high acquisition cost, the past well-publicized system failures, and loss of prestige.

The importance of quality in business and industry is increasing rapidly because of factors such as competition, growing demand from customers for better quality, increasing number of quality-related lawsuits, and the global economy. Nonetheless, the cost of quality control accounts for around 7-10% of the total sales revenue of manufacturers. Today, companies are faced with reducing this amount and at the same time improving the quality of products and services for their survival in the internet economy.

Today, safety has become a critical issue because each year a very large number of people die and get seriously injured due to workplace and other accidents. For example, in the U.S according to the National Safety Council(NSC), there were 93,400 deaths and a vast number of disabling injures due to accidents in 1996. The total cost of these accidents was estimated to be around $121 billion. Some of the other factors that are also playing an instrumental role in demanding the need for better safety are government regulations, public pressures, and increasing number of lawsuits.

Terminology

safety
quality
space satellites
reliability

Today, billions of dollars are being spent annually to produce new products using modern technologies. Many of these products are highly sophisticated and contain millions of parts. For example, a Boeing 747 jumbo jet is made up of approximately 4.5 million parts including fasteners. Needless to say, reliability, quality, and safety of systems such as this have become more important than ever before. Global competition and other factors are forcing manufactures to produce highly reliable, good quality, and safe products.

It means that there is a definite need for the reliability, quality, and safety professionals to work closely during design and other phases. To achieve this goal it is essential that they have to a certain degree an understanding of each other's discipline. Once each of these professionals has the necessary knowledge of each other's discipline, many of their work-related difficulties will disappear or at least be reduced to a tolerable level; thus resulting in better reliability, quality, and safety of end products or systems.

EXERCIS 15-

A. Words and Phrases

1. lawsuit
2. play a key role in
3. complexity
4. competition
5. acquisition
6. prestige
7. nonetheless
8. revenue
9. regulation
10. needless to say
11. reliable
12. sophistication
13. account for

B. Reading Comprehension

1. Put T or F to indicate if the sentences below are true or false according to the facts in the passage.

- The cost of quality control accounts for around 5% of the total sales revenue of manufacturers. ________
- The only factor that makes reliability important is the number of lawsuits. ________
- Global competition and other factors make manufacturer produce highly reliable, good quality, and safe products. ________
- The reliability, quality, and safety professionals must understand each other's discipline. ________
- Billions of dollars are being spent annually to prevent car accidents.
- Today, companies are trying to increase the budget for the cost of quality control. ________
- In the U.S. there were more than 100 thousands deaths due to accidents in 1996. ________

2. Answer the following questions.

- Explain the reason why the quality in business and industry becomes important?
- How many parts are needed to assemble a Boeing 747 jumbo jet?
- How much is the total cost of accidents occurred in the U.S. in 1996?
- List some of the specific factors that are playing a key role in increasing the importance of reliability in designed systems.
- What other factors besides cost play an instrumental role in demanding the need for better safety?

bibliography

B.S Dhillon, *Reliability, quality, and Safety for Engineers*, pp. 1~2, CRC Press, 2005.

Unit

16

Mechanical Design and MEMS

Unit 16 Mechanical Design and MEMS

1) Mechanical Design Process

The process of design can be schematically represented to levels of increasing formality and complexity as shown in Figs. 16.1 and 16.2.

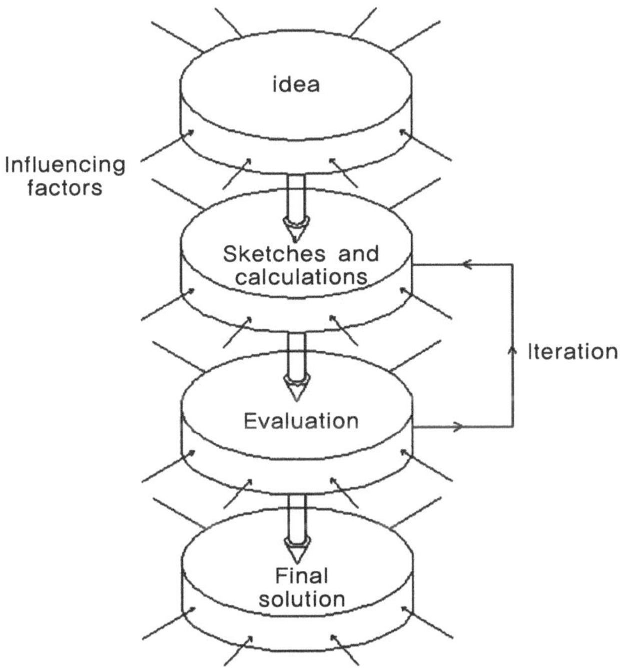

▲ **Fig 16.1** The traditional and familiar 'inventor's' approach to design.

◎ **Terminology**

schematically

Fig 16.1 represents the traditional approach associated with lone inventors comprising the generation of the 'bright idea', drawings and calculations giving form to the idea, judgement of the design and re-evaluation if necessary, resulting in the generation of the end product. Fig 16.2 shows a more formal description of the design process which might be associated with engineers operating within a company's management structure. The various terms used in Fig 16.2 are described below.

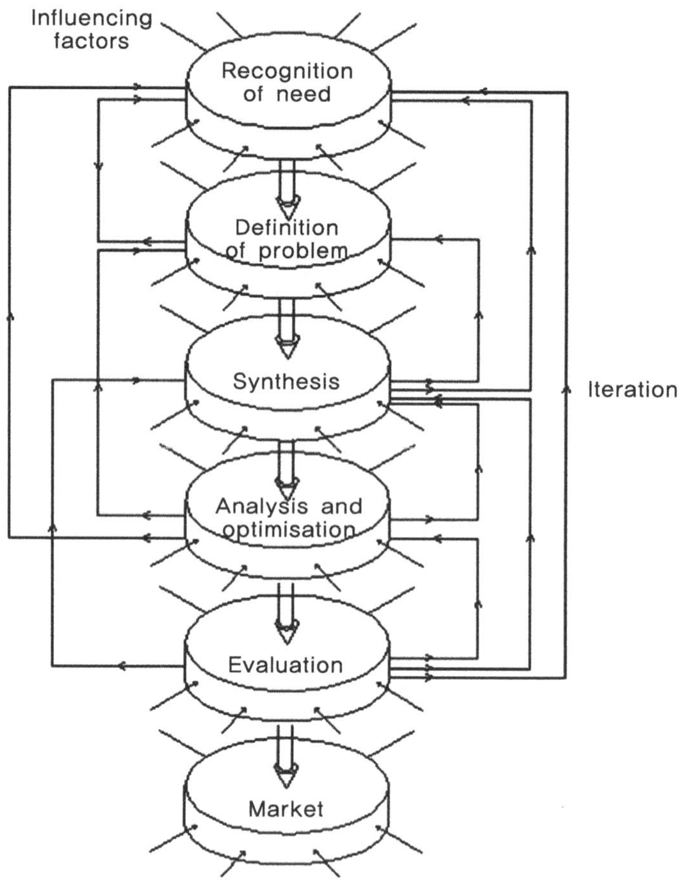

▲ Fig 16.2 The design process

- Recognition of need

Often design begins when an individual or company recognizes a need, or identifies a potential market, for a product, device or process. Alternatively 'need' can be defined as when a company decides to reengineer one of its existing products(for example producing a new car model).

- Definition of problem

This involves all the specification of the product or process to be designed. For example, this would include inputs and outputs, characteristics, dimensions and limitations on quantities.

- Synthesis

This is the process of combining the ideas developed into a form or concept which offers a potential solution to the design requirement. The term synthesis may be familiar from its use in chemistry where it is used to describe the process of producing a compound by a series of reactions of other substances.

- Analysis

This involves the application of engineering science — subjects explored extensively in traditional engineering courses such as statics and dynamics, mechanics of materials, fluid flow and heat transfer. These engineering 'tools' and techniques can be used to examine the design to give quantitative information such as whether it is strong enough or will operate at an acceptable temperature. Analysis and synthesis invariably go together. Synthesis means putting something together and analysis means resolving something into its constituent parts or taking it to pieces. Designers have to synthesise something before it can be analysed — the famous chicken-and-egg scenario! When a product is analysed, some kind of deficiency or inadequacy may be identified requiring the synthesis of a new solution prior to reanalysis and repetition of the process until an adequate solution is obtained.

Terminology

heat transfer
synthesis
fluid flow
mechanics of materials
potential solution
statics
dynamics
reengineer

- Optimisation

This is the process of repetitively refining a set of often conflicting criteria to achieve the best compromise.

- Evaluation

This is the process of identifying whether the design satisfies the original requirements. It may involve assessment of the analysis, prototype testing and market research.

Although Figs 16.1 and 16.2 at first sight show design occurring in a sequential fashion, with one task following another, the design process may actually occur in a step-forward, step-back fashion. For instance, you may propose a solution to the design need and then perform some calculations or judgements which indicate that the proposal is inappropriate. A new solution will need to be put forward and further assessments made. This is known as the iterative process of design and forms an essential part of refining and improving the product proposal.

Terminology

prototype testing
iterative process
criteria
design methodology

Note that the flow charts shown in Fig 16.1 and 16.2 do not represent a method of design but rather a description of what actually occurs within the process of design. The method of design used is often unique to the engineer or design team. Design methodology is not an exact science and there are indeed no guaranteed methods of design. Some designers work in a progressive fashion; others work on several aspects simultaneously.

EXERCIS 16–

A. Words and Phrases

1. formality	2. complexity	3. comprising
4. evaluation	5. alternatively	6. specification
7. extensively	8. traditional	9. quantitative
10. invariably	11. constituent	12. scenario
13. deficiency	14. inadequacy	15. repetition
16. conflicting	17. compromise	18. sequential
19. inappropriate	20. assessment	21. progressive
22. aspect	23. at first sight	

B. Reading Comprehension

1. Put T or F to indicate if the sentences below are true or false according to the facts in the passage.

- The first step in a formal design process is 'definition of problem'. ________________
- The process of identifying whether the design satisfies the original requirements is called 'optimization'. ________________
- Design methodology is an exact science and is always guaranteed. ________________
- 'Analysis' stage involves the application of engineering science. ________________
- Synthesis means putting something together. ________________
- The process of repetitively refining a set of often conflicting criteria to achieve the best compromise is 'evaluation'. ________________
- Analysis and synthesis invariably go together. ________________

2. Answer the following questions.

- What was the first stage of the design process in the generation of the 'bright idea'?
- What is the final step of the recent design process?
- How is the term 'need' defined in this paragraph?
- Explain 'synthesis' in contrast to 'analysis'. How are they different?

bibliography

Peter R.N. Childs, *Mechanical Design*, pp. 1~2, Arnold, 1998.

2 MEMS and Nanotechnology

Nano- and microelectromechanics, optoelectronics, as well as nano- and microfabrication have accomplished phenomenal growth over the past few years due to rapid advances in theoretical developments, experimental results(using state-of-the-art measurement and instrumentation hardware), high-performance computer-aided design software and computationally efficient environments. Recent fundamental and applied research and developments in nano- and microelectromechanics informatics, and nanotechnolohy have notably contributed to the current progress. Leading edge basic research, novel technologies, software, and hardware are integrated to devise new systems and study existing systems. These prominent trends provided researchers, engineers, and students with the needed concurrent design of integrated complex microelectromechanical and nanoelectromechanical systems(MEMS and NEMS). The synergy of engineering, science, and technology is essential to attaining the goals and objectives. It becomes increasingly difficult to perform analysis and design of micro- and nanoscale systems, subsystems, devices and structures without the unified theme and multidisciplinary synergy due to integration of new phenomena, complex processes, and compatibility. Therefore, micro- and nanoelectromechanics theories must be further developed and applied micro- and nanoelectromechanics are based upon fundamental theory, engineering practice, and leading edge technologies in fabrication of micro- and nanoscale systems, subsystems, devices, and structures which have dimensions of micrometers and nanometers.

The development and deployment of MEMS and NEMS are critical to the U.S. economy and society because nano- and microengineering and science will lead to major breakthroughs in information technology, computers, medicine, health, manufacturing, transportation, energy, avionics, security,

Terminology

informatics
electromechanics
leading edge
nanoscale
optoelectronics
NEMS
multidisciplinary
microfabrication
nanotechnolohy
MEMS
micro

etc. For example, MEMS and NEMS have an important impact on medicine and bioengineering (DNA and genetic code analysis and synthesis, drug delivery, diagnostics and imaging), bio and information technology, avionics and aerospace(nano- and microscale actuators and sensors, smart reconfigurable geometry wings and blades, space-based flexible structures, microgyroscopes), automotive systems and transportation(transducers and accelerometers), manufacturing and fabrication(micro-and nanoscale smart robots), safety, etc. Therefore, MEMS and NEMS will have tremendous positive direct and indirect social and economic impacts. New MEMS and NEMS will be divised and used in a variety of applications(electronics, medicine, metrology, etc). Micro- and nanoscale structures, devices, and systems frequently have the specific applications such as high-frequency resonators, electromagnetic field and stress sensors, etc.

Nanoscale devices and systems will allow one to access the atomic scale phenomena, where quantum effects are predominant. Though, depending upon the size and architecture, MEMS can be designed using scaling paradigm as the microstructures dimension is within 0.1 μm, the secondary effects cannot be neglected for nanostructures. Hence, advanced high fidelity modeling and analysis concepts are used for various MEMS. NEMS usually are not scalable which leads to numerous challenges. To support the nano- and microengineering, basic, applied, and experimental research as well as engineering developments must be performed.

Nanoengineering studies nanoscale structures, devices and systems, whose structures and components exhibit novel physical(electromagnetic, electromechanical, optical, etc.), chemical, electrochemical, and biological properties, phenomena, and effects. The dimension of NEMS and their components(nanostructures) is from 10^{-10} m(molecule size) to 10^{-7} m; that is, from 0.1 to 100 nanometers. To manufacture nanostructures. nanotechnology is applied. In contrast, the dimension of MEMS and their components (devises and microstructure) is from 100 nanometers to the

Terminology

electromagnetic
transducer
DNA
genetic code
atomic scale
avionics
gyroscope
resonator
accelerometer
quantum effect
metrology
Nanoengineering
microstructure
bioengineering
nanostructure

centimeter range.

Conventional microelectronics technologies are usually used to fabricate MEMS. Studying nanoscale systems, one concentrates on the atomic and molecular levels in manufacturing, fabrication, design, analysis, optimization, integration, synthesis, etc. Reducing the dimension of systems leads to the application of novel materials(carbon nanotubes, molecular wires, etc) and new fabrication technologies. The problems to be solved range from high-performance MEMS and NEMS. For example, micro- and nanoscale switches, logic gates, actuators, and sensors should be devised, studied, optimized, and fabricated.

All living biological systems function as a result of atomic and molecular interactions. Different biological and organic systems, subsystems, and structures are built. The molecular building blocks(proteins, nucleic acids, lipids, carbohydrates, DNA and RNA) are applied to perform biomimicking and prototyping to devise novel MEMS and NEMS. The devised MEMS and NEMS must have the desired properties, characteristics, and functionality. Analytical and numerical methods must be developed in order to analyze the dynamics, three-dimensional geometry, bonding, and other features of atoms and molecules. In order to accomplish this analysis, complex electro-magnetic, mechanical, and other physical and chemical properties must be studied.

Micro- and nanosystems will be widely used in medicine and health. Among possible applications are drug synthesis, drug delivery, nanosurgery, nanotherapy, genome synthesis, diagnostics, novel actuators and sensors, disease diagnosis and prevention, nonrejectable artificial organs design and implant, biocompatible materials and structures, etc. For example with drug delivery, the therapeutic potential will be enormously enhanced due to direct effective delivery of new types of drugs to the specified body sites. The molecular building blocks of DNA-based structures, proteins, nucleic acids, lipids, as well as carbohydrates and their non-biological mimics are examples

Terminology

carbohydrate
RNA
molecular wire
molecular
protein
atomic
carbon nanotube
organic system
nucleic acid
logic gate
actuator
lipid
carbohydrate
implant

of materials that possess unique properties determined by their size, folding, and patterns at the nano- and microscale. Significant progress has been made, and new analytical tools must be developed capable of characterizing the chemical, electrical, and mechanical properties of cells including processes such as cell division, locomotion, and propulsion.

Microelectromechanical systems, which integrate motion microdevices (actuators and sensors), radiating energy microdevices(antennas, microstructures with windings), microscale driving/sensing circuitry, and controlling / processing integrated circuits(ICs), are widely used.

Terminology

cell division
analytical tool

EXERCIS 16–

A. Words and Phrases

1. phenomenal
2. theoretical
3. prominent
4. synergy
5. compatibility
6. breakthrough
7. diagnostics
8. reconfigurable
9. predominant
10. fidelity
11. interaction
12. biomimicking
13. prototyping
14. implant
15. biocompatible
16. therapeutic
17. circuitry
18. deployment
19. deployment
20. avionics
21. paradigm
22. biomimic

B. Reading Comprehension

1. Put T or F to indicate if the sentences below are true or false according to the facts in the passage.

- MEMS does not have anything to do with medicine and bioengineering. ________
- MEMS has positive social and economic impacts. ________
- It is necessary to use nanoscale devices to access the atomic scale phenomena. ________
- The dimension of NEMS is from 10^{-12} m to 10^{-10} m. ________
- Atomic and molecular interactions make all living biological systems function. ________
- For fabrication of MEMS, conventional microelectronics technologies cannot be used. ________

2. Answer the following questions.

- List the examples of materials that possess unique properties determined by their size, folding at nano- and microscale in drug delovery.
- What does the term 'NEMS' means? Rewrite in full name.

bibliography

Sergey Edward Lyshevski, *MEMS and NEMS*, pp. 1~3 CRC Press, 2002.

수정 Engineering English
FOR MECHANICAL ENGINEERS

2009년 2월 25일 초판 인쇄
2016년 8월 31일 4쇄 발행

저 자 김동혁, 유삼상, 이건수, 도덕희
발 행 인 송기수
발 행 처 도서출판 GS인터비전
편 집 처 도서출판 GS인터비전
인 쇄 처 대일 문예사
등록번호 제 313-2011-120호
I S B N 978-89-93668-10-0 (92530)

주 소 서울 은평구 증산로15길 69, 2층
전 화 02-976-7898, 3272-7898, 0707-720-7898
팩 스 02-6468-7898
E-Mail gsinter7@gmail.com
홈페이지 gsintervision.co.kr

정 가 15,000원